SPIN GLASSES
AND BIOLOGY

SERIES ON DIRECTIONS IN CONDENSED MATTER PHYSICS

Directions in Condensed Matter Physics – Vol. 6

SPIN GLASSES
AND BIOLOGY

Edited by

Daniel L Stein

Department of Physics
University of Arizona

 **World Scientific**
Singapore • New Jersey • London • Hong Kong

Published by

World Scientific Publishing Co. Pte. Ltd.

P O Box 128, Farrer Road, Singapore 912805

USA office: Suite 1B, 1060 Main Street, River Edge, NJ 07661

UK office: 57 Shelton Street, Covent Garden, London WC2H 9HE

British Library Cataloguing-in-Publication Data
A catalogue record for this book is available from the British Library.

First published 1992
First reprint 1998

SPIN GLASSES AND BIOLOGY —
Directions in Condensed Matter Physics Vol. 6

ISBN 9971-50-537-1
ISBN 9971-50-538-X (pbk)

Printed in Singapore by JCS Office Services & Supplies Pte Ltd

Preface

Forays of physicists into areas outside of the traditional boundaries of their discipline are nothing new, but their number virtually exploded throughout the 1980's. A flourishing subset of these pursuits includes the many attempts to connect ideas developed in the study of spin glasses to problems arising from or associated (however vaguely) with biology.

As Philip Anderson points out in the introduction, the modern study of spin glasses proper is at least two decades old, but our present understanding of some very basic issues is nonetheless still quite limited. Two valid objections to the application of spin glass physics to biological (or biology-related) problems naturally follow. The first involves the transfer of concepts developed for a system that is only partially understood. What is at issue here, though, is our understanding of the concept, not the system, and the appropriateness of its use in studying a particular problem. This is the basic thrust of the articles which appear within, and in this sense a better title for the book might have been "Spin Glass Notions and Biology".

The second objection involves the potential for (scientific) catastrophe accompanying any attempt to apply a physical model to a biological system. Experience shows that the physics-prone use in biology of Occam's razor is more than likely to lead to self-inflicted wounds. Consequently, if the spin glass portion of the title needs to be taken less than literally, then even more so for the biology part. While some of the articles do deal directly with biological problems — for example, protein folding — others deal with issues, abstract or concrete, which arise from the study of biological systems. One such example is what has come to be called "biological computation". Moreover, the flow of ideas has worked equally well in the opposite direction — John Hopfield's early connectionist models of neural networks provided the physics community with new statistical mechanical models possessing quenched disorder, whose study has been interesting in and of itself.

Among the various pitfalls which confront the physicist attempting to apply a physical concept or model to a biological problem is that of trying to force the square peg into the round hole. Any system needs to be dealt with in its own right, its phenomenology understood and accounted for. Certainly no physics model can be applied to biology in a wholesale fashion. Each physics foray into biology should be carefully circumscribed, its limitations carefully laid out, and the particular *features* being studied (as opposed to the system in its entirely) made explicit. Even the simplest protein, for example, is far more complicated than any statistical mechanical system which physicists attempt to study; structure, dynamics, and function are often tightly interwoven. Application of spin glass ideas here might seem doubly strange, considering that from the point of view of evolution and function the protein is hardly a random object. But such sweeping statements as "the protein is a spin glass" completely miss the point of the research they refer to. It would indeed be folly to try to understand the entire protein all at once, and certainly not within the context of a single physical model, but narrow focus on one or a few *specific* properties, coupled with a judicious application of appropriate ideas, might yield some enlightenment. This is what is attempted throughout this book, and unquestionably some progress in this direction has been made.

What *are* the notions from spin glass physics that have been used in biological (and other) applications? Most are discussed cogently in the introduction by Anderson (who introduced, or played a role in the creation of, many of them): frustration, quenched disorder, replicas, ultrametricity, metastability, irreversibility, and maybe robust criticality, among others. At the time this is being written (November 1991), the fields of spin glass physics proper and that of applying it to problems from other disciplines have diverged somewhat, the latter taking on a life of its own. Because this last is just the subject of the present volume, I will confine the present discussion to a brief survey of the former.

By the mid-1980's, it was generally agreed upon that the statistical mechanics of the infinite-ranged Ising spin glass (or the Sherrington-Kirkpatrick[1] model) was mostly understood. The replica-symmetry breaking scheme discovered by Parisi[2] exhibited an infinite number of stable[3] low-temperature states whose properties were in good agreement with numerical simulations. But the nature of ergodicity-breaking was rather surprising: the distances among these states in configuration space had an underlying ultrametric structure,[4] i.e., the distances were related via a type of hierarchical tree. Whether more realistic short-ranged (Edwards-Anderson[5]) models also possessed this property immediately became an important question.

This was answered in the negative by proponents of a scaling *ansatz* for the nearest-neighbor Edwards-Anderson Ising model.[6-8] These theories suggested that, unlike the infinite-ranged model, no more than two pure states (related via a global flip) exist at any temperature and field.[9] The argument between those who favor an SK-like picture for short-ranged, low-dimensional spin glasses and those who advocate the scaling scenario continues unresolved, and in particular the important question of multiplicity of pure states (or ground states at zero temperature) remains open.

The preceding discussion focuses on only one sets of issues related to Edwards-Anderson Ising spin glasses. In fact, the Ising Edwards-Anderson spin glass constitutes only a small subset of possible spin glass models, but has tended to dominate theoretical work, both analytical and numerical, on (what are presumed to be) realistic spin glasses. The question of multiplicity of pure states in this model is primarily a theoretical one, although important; other outstanding issues include proving the presence (or absence) of a phase transition in any "realistic" model, elucidating the nature of the low-temperature phase(s), understanding the effects of an external magnetic field, and the like. In order to better compare theory with experiment, it is important to understand the similarities and differences of the critical and low-temperature properties of different spin glass Hamiltonians: in particular, nearest-neighbor models, models with random long- (but not infinite-) ranged couplings (say, falling off as $1/r^d$ in d dimensions), and randomly site-diluted RKKY models, among others.

These comprise only a small subset of questions which both theorists and experimentalists would like to see answered; I mention them here because, while they need to be resolved before we can claim to have even a modicum of understanding of spin glasses, their resolution is likely to have little or no effect on biological applications of the type discussed in this book. (It is important to remember that the issue of multiplicity of pure states does not bear on the issue of the existence of many locally stable, or metastable, states, which may indeed be important in biological applications.) It is often stated, in this context, that results of the kind discussed above apply strictly to infinite systems, while biological problems require a consideration of finite systems. This objection certainly applies some of the time, but probably less often than is usually thought. More importantly, the kinds of problems discussed above, crucial as they are for understanding physical spin glasses, simply bear no obvious relevance to biological problems.

Of course, exceptions can always be found: understanding the low-temperature spin glass phase or phases may be very useful, and even multiplicity of pure states does bear some relevance to the number of stable memories in some neural network models, for example. But more to the point, there are many other open problems (in fact, almost any problem relating to non-infinite-ranged spin glasses is open), and several are of more direct importance for further progress in applying spin glass physics to biology. In particular, any advances in understanding the dynamics of realistic spin glasses will be extremely helpful. Anderson's ideas on robust criticality, and possible connections to the ideas of Packard, Kauffman, *et al.*, provide an intriguing direction for new research. Even at a more basic level, the connections between different aspects of biological systems and certain prominent features of spin glasses, including frustration, metastability, and irreversibility, have barely been explored. The application of results arising from other spin glass models, such as Derrida's random energy model or the Potts glass, has also provided significant research pathways. Progress in these areas and in biology-related investigations are strongly coupled, and we can reasonably expect the flow of ideas to travel in both directions. It is our hope that this book can contribute to the furtherance of these objectives.

About the book

The focus of this volume is to present some of the major thrusts of current research using spin glass ideas in studying biological or related problems. An exhaustive compilation is not attempted; rather, a representative sample of some of the main streams of this type of research is presented in order to convey some of the flavor and variety of the field. No essays on spin glasses proper appear; the reader is assumed to have some familiarity with the subject, although a reasonable amount of spin glass physics may be picked up from some of the articles in this book. Several excellent reviews and books can be consulted for those wishing to learn about spin glasses themselves; some of the more recent include that exhaustive 1986 review of Binder and Young,[10] a much shorter and far-from-exhaustive review by myself,[11] and the books by Mezard *et al.*,[12] Chowdhury,[13] and Hertz and Fischer.[14] At a very basic and qualitative level there is also my Scientific American article on spin glasses and applications.[15]

The articles within cover a wide-ranging set of topics, including neural networks, biological evolution, the immune response, protein dynamics and folding, and general models of adaptation. We briefly discuss each in turn.

The article by Hanoch Gutfreund and Gérard Toulouse on neural networks traces the introduction of statistical mechanical ideas into the study of biological computation, cognition, and learning. Simplified models of the neuron, basic network architectures, and dynamical rules are introduced. This sets the stage for a discussion of how these systems exhibit features such as associative memory, learning, categorization, and generalization. The Hopfield model, which heralded the arrival of spin glass ideas into the arena of neural computation, is discussed in detail, along with its variations. The important ideas of Elizabeth Gardner and the work of Gardner and Bernard Derrida are introduced and discussed, followed by an extensive treatment of hierarchical data structures. The problems of learning and generalization are then treated, and the article concludes with a discussion of outstanding issues, including that of biological relevance.

The following four articles relate, in one way or another, to the central problem of adaptation. Organisms competing in a given ecosystem comprise the most familiar example, but others abound as well, from prebiotic selection to maturation of the immune response. Spin glass physics, particularly the feature of the "rugged landscape", is ideally suited for studying various features of these types of problems, and a major portion of this book is devoted to such explorations. The chapter by Stuart Kauffman initiates a general discussion of random walks on rugged landscapes, which provides a unifying focus for subsequent treatments of adaptation (from the spin glass point of view). The now venerable ideas of Sewell Wright on fitness landscapes is recast into spin glass language via Kauffman's Nk model. Applications to various problem in evolution in its broadest sense (including, for example, the immune response) are discussed, but the chapter focuses mainly on elucidating the general features of such walks: numbers of local optima, average number of steps to the nearest optimum, sizes of basins of attraction of optima, and so on. The chapter closes with a discussion of what Kauffman calls the "complexity catastrophe", tunable families of landscapes, and measuring correlation structures of rugged landscapes.

Daniel Rokhsar next treats the specific case of prebiotic evolution, or more precisely the emergence of biological information encapsulated within a molecular framework. The work discussed here originated from a 1983 PNAS paper by Anderson[16] and was further developed by Anderson,

Rokhsar, and myself, with more recent progress due to Albert Wong.[17,18] Rokhsar begins with a general discussion of biological information, in particular what it means and what its existence requires. Simple models of evolution of information are then presented, including Manfred Eigen's important contribution of the hypercycle. Anderson's introduction of a spin-glass-like "death function" is next discussed, and a simple picture containing all of the foregoing elements is put together. Numerical simulations on the resulting model are presented, and the kinds of behavior which emerge are discussed and placed within the overall context of the chapter.

Gérard Weisbuch's chapter is concerned with punctuated evolution. The relation of phenotype and genotype within an evolutionary context is introduced, followed by a discussion on population genetics. Weisbuch presents a genetic model based on networks of automata based on earlier ideas of Kauffman, and investigates its resulting properties, such as robustness, sensitivity to initial conditions, and emergence of patterns. The model is then used to offer a resolution of some problems related to punctuated equilibria in evolution, and once again the characteristics of the rugged landscape become central.

The chapter by Alan Perelson and Catherine Macken also employs the physics of rugged landscapes, this time in an application to the maturation of the immune response. The authors present a simple set of models (which have the virtue that they can be studied analytically) of affinity maturation, and show how a direct application of the ideas discussed in Kauffman's chapter can be used to determine their properties.

The final two chapters represent a different approach is utilizing spin glass ideas in biology. The chapter of Robert Austin and Christine Chen, written in Bob's inimitable style, discusses the host of ideas and the problems arising from the application of spin glass physics to conformational structure, thermodynamic properties, and dynamics of globular proteins in the native state. Myoglobin is the primary workhorse, though hemoglobin and calmodulin are also discussed at various stages. Theoretical ideas are introduced throughout the chapter with a strong concern for applicability to experiment. Both the successes and limitations of these ideas are discussed at length, and a good sense is conveyed of the difficulty of extracting relevant information from experiments on a system as horribly complicated as even the simplest protein. Therefore, any comparison of theory with experiment must be made with the utmost caution.

The concluding chapter by Peter Wolynes also concerns itself with proteins, but in a wholly different arena: the problem of folding. This article represents the most recent major application of spin glass ideas to biological problems, and is one of the most exciting not only because of the significance of the problem it attempts to address, but also because of the possibilities of spin glass "engineering" it suggests. Progress in this area and others, such as neural networks, are strongly correlated.

Acknowledgements

This volume is the result of the efforts of many; the Scientific Advisory Board of World Scientific Publishing conceived of the idea for the book, and the staff, particularly Ms. P. H. Tham, expended a good deal of effort preventing me from being wholly negligent in my editorial duties. T.-V. Ramakrishnan was primarily responsible for securing my participation in the project, and I hope someday to return the favor. I extend my sincere thanks to all involved, especially the contributors to the volume, all of whom have played a prominent role in the development of the application of spin glass ideas to biology.

November 1991   D. L. Stein
                                                                             Tucson, Arizona

References

1. D. Sherrington and S. Kirkpatrick, *Phys. Rev. Lett.* **35**, 1972 (1975).
2. G. Parisi, *Phys. Rev. Lett.* **43**, 1754 (1979); G. Parisi, *Phys. Rev. Lett.* **50**, 1946 (1983).
3. C. De Dominicis and I. Kondor, *Phys. Rev.* **B27**, 606 (1983).
4. M. Mezard, G. Parisi, N. Sourlas, G. Toulouse, and M. Virasoro, *Phys. Rev. Lett.* **52**, 1156 (1984).
5. S. Edwards and P. W. Anderson, *J. Phys.* **F5**, 965 (1975).
6. W. L. McMillan, *J. Phys.* **C17**, 3179 (1984).
7. A. J. Bray and M. A. Moore, in *Heidelberg Colloquium in Glassy Dynamics*, ed. J. L. van Hemmen and I. Morgenstern (Springer-Verlag, 1987).
8. D. S. Fisher and D. A. Huse, *Phys. Rev. Lett.* **56**, 1601 (1986); D. S. Fisher and D. A. Huse, *Phys. Rev.* **D38**, 386 (1988).
9. D. A. Huse and D. S. Fisher, *J. Phys.* **A20**, L997 (1987); D. S. Fisher and D. A. Huse, *J. Phys.* **A20**, L1005 (1987).
10. K. Binder and A. P. Young, *Rev. Mod. Phys.* **58**, 801 (1986).
11. D. L. Stein, in *Lectures in the Sciences of Complexity*, ed. D. L. Stein (Addision-Wesley, 1989), pp. 301–353.
12. M. Mézard, G. Parisi, and M. Virasoro, *Spin Glass Theory and Beyond* (World Scientific, 1986).
13. D. Chowdhury, *Spin Glasses and Other Frustrated Sytstems* (Wiley, 1986).
14. J. A. Hertz and K. H. Fischer, *Spin Glasses* (Cambridge University Press, 1991).
15. D. L. Stein, *Scientific American* **260**, 52 (1989).
16. P. W. Anderson, *Proc. Natl. Acad. Sci. USA* **80**, 3386 (1983).
17. A. J. Wong, *J. Theor. Biol.* **146**, 523 (1990).
18. A. J. Wong, in *1990 Lectures in Complex Systems*, ed. L. Nadel and D. L. Stein (Addison-Wesley, 1991), pp. 547–553.

Contents

Introduction

P. W. Anderson

Joseph Henry Laboratories of Physics, Jadwin Hall, Princeton University, Princeton, NJ 08544, USA

Two seminal events for the subject of this book happened in the year 1969: the first paper entitled "spin glass" was published[1] in which I borrowed the term "magnetic glass" from "cet Mauvais Gaulois"[2] by Bryan Coles and shortened it; and Stuart Kauffman proposed a Boolean network computer model for evolution.[3] The first mode was random and — in a primitive fashion — non-ergodic, but not of course biological; the second was not yet random nor non-ergodic but biological. Much the same period marks the beginnings of complex neutral network theory with ideas due to Cooper, Little, Rumelhart and others. These three strands of theory did not begin to interact until barely 10 years ago, yet an examination of the bibliographies of the various chapters to this book will reveal an already rich literature. I suppose, if nothing is done to prevent it, that some enterprising editor can be expected to start a journal on spin glasses, as in fact one already exists on neutral networks.

The hybridization of fields which has inspired this outpouring of papers and even of ideas had to wait another 10 years, at least, and came via contacts between a rather small number of individuals in the early 80's, many of whom are represented in this book. Regrettably, John Hopfield is not, although his paper[4] on a spin glass model of associative memory was the single discrete event which most nearly epitomizes, and more or less began, this period. His ideas are well described by Toulouse and Gutfreund in their chapters.

The "heroic period" of spin-glass theory, from Edwards-Anderson to Toulouse and Parisi, Virasoro, and Mézard ran from 1975–1980; at the end of that time we had a good understanding of the long-range Sherrington-

1

Kirkpatrick model in terms of a brand new statistical mechanics, and a hope that the same scheme worked for other cases. Hopfield knew of these ideas via our contact at Bell Labs., but nonetheless his was brilliant leap of imagination to the idea of the neural network as a programmable spin glass. Meanwhile, my long-standing interest in early evolution led to modelling attempts which, in 1981, during visits to Hopfield and Orgel, I realized needed a rugged spin-glass landscape to evolve in.[5]

Several meeting places seem to have played a role in the further spread of this contagion. A workshop at Aspen brought Hans Kuhn, Hopfield, Gérard Weisbuch, Toulouse, Stein, myself, and probably others I have forgotten, together. Gene Yates, in Dubrovnik 1980, brought together Steve Gould, Harold Morowitz, Leslie Orgel, myself and Schuster, and we shared ideas on early evolution.[6] The unlikely auspices of Werner Erhard and EST brought together Kauffman, Packard, myself, and I think Hopfield in San Francisco in this period. But most important of all, the founding workshops of the Santa Fe Institute in 1984–86 found almost all of the cast of characters in place, including now Manfred Eigen, Miguel Virasoro and Gérard Toulouse as well.[7]

Dan Stein, working next door to me, and a collaborator on evolution theory, was the father of the further applications to protein dynamics,[8] taken up enthusiastically by others as this book testifies. This in turn served as the springboard for more recent work on protein folding.

Enough for history; what will you find in the book, and what further advances may happen?

"Spin glass", in a generalized sense, is a short-hand for "quenched random", "frustrated" interactions among a large set of nearly identical, simple objects — such as single bases of DNA seen as the characters in a string, such as neuron models, such as amino acid orientations, or perhaps even more general objects such as genes. "Quenched" means fixed, not to be altered at least in the short-term dynamics; "frustrated" means that no trivial single configuration will satisfy all interactions simultaneously. A "large number" means that a meaningful "large N" or "thermodynamics limit" is envisaged. All existed in the original physical spin glasses, alloys of Mn or Fe in Cu, Ag, or Au.

The above mentioned "heroic period" left us with a set of concepts and techniques for dealing with the unique statistical mechanics of this kind of system: the concepts of the "spin glass order parameter" representing a long-range order in time rather than space, of "frustration" and of "ultrametric sets of solutions" rather than of a simple, single equilibrium state; and of "broken ergodicity"; and with the techniques of the "replica" and "cavity" (TAP) methods which deal respectively with sets of solutions and with the individual solution, but as one typical solution of a non-linear random matrix problem.

Each of the articles uses this apparatus in different way. (I do not distinguish the "N-k" realization of frustrated interactions used by Kauffman and Weisbuch as clearly different in principle.) In neural net theory, the brain is pre-programmed to arrive at different members of the solution set depending on different inputs; in evolution theory, the fitness of populations drives a continuous search for a best available — any best available — solution; in protein theory, we worry about how to get to a unique solution — and how unique is it? We also expect to use the sensitivity of the spin-glass susceptibility to explain configuration changes and catalytic activity.

Is the refined spin glass mathematics essential in biology? Probably not at all, but like all such overarching concepts, it seems to this observer that seeing the whole of the forest — i.e., the ecology — may be at least as important as staring at one tree or one leaf. The possibility of analytic or modelling treatments of complicated problems like these invariably leads to new insights in each individual example.

Where else can we go? I think a long way. Stu Kauffman will already hint that *interacting* spin-glasses may have a general, exciting evolutionary dynamics, for instance, leading to "adaptation to the edge of chaos". In neural net theory, Miss Gardner's great insight of setting the solutions and varying the interactions opens whole new worlds for us and helps free us from the limitation to symmetric connections.

Even further in the future, we may note that Mézard and Parisi, and Georges, Mézard and Yedidia,[9] have recently broken the problem of finite-range quenched randomness. What new ideas this may lead to in biology are not clear.

Another little-appreciated feature of spin glass theory is that spin glass is the only quasi-equilibrium system which is, below its T_c, *always in a critical state, regardless of tuning*. The pervasiveness of scale-free, fractal behavior in nature calls out for an explanation in terms of criticality, as Bak *et al.* suggested; but here we have a new way of finding *robust criticality*. Whether spin glass-like systems will serve as a description or as a source of criticality and scaling laws in nature remains to be seen, but looks very hopeful.

Therefore we do not know whether this book will mark the end of the immersion of statistical mechanics into biology, or the beginning of a new and even more exciting period. My own belief is that the latter is true. The statistical theory of life cannot be an equilibrium or quasi-equilibrium theory of unique equilibrium outcomes or unique "dissipative structures", because no aspect of life is unique and inevitable; but life requires some organizing principle, some sorting out of configuration space into stable attractors for which the broken ergodicity principle and the ultrametric solution tree of the spin glass provide an ideal theoretical framework. Spin glass will never cure boils or feed a cow, but if general understanding of how life can actually have come to be the way it is and of how the whole structure works is one's goal, there is no alternative conceptual structure. This "untuned criticality" is behind, also, the fact — already mentioned — that the protein can change configuration macroscopically when a single bond is changed, (that is, it can "avalanche") just as it is behind the "punctuated equilibria" of evolution theory.

It was a dream of many people, starting clearly with Onsager,[10] that eventually a genuine statistical mechanics of non-equilibrium systems would rise. Turing, Prigogine, Kuhn and Eigen dreamed that such statistical mechanisms could point the way to some general theory or concept of life and how it arose. The earliest attempts were clearly not satisfactory. The "dissipative structures" of pattern formation are never robust, while deterministic chaos as a model is not a statistical mechanics but a theory of systems with very few relevant degrees of freedom. On the other hand life's dominant characteristic is the steady complexification of the space of living things.

We may be lurching towards a new attempt which has more chance to be successful, and is embodied in three general principles, each controversial, none of them truly well posed, yet infinitely hopeful:

1. The broken ergodicity and solution tree concepts of the spin glass;
2. Self-organized criticality, avalanches, and scale-free behavior in now at least two types of models;
3. The concept of adaptation to the edge of chaos of Packard and Kauffman.

This book leads us along the first steps on this ambitious path.

References

1. P. W. Anderson, *Mat. Res. Bull.* **5**, 549 (1970).
2. G. Toulouse, in historical remarks quoted in *Spin Glass Theory and Beyond* (World Scientific, 1986). The translation, of course, is "that mischievous Welshman".
3. S. Kauffman, *Nature* **224**, 177 (1969).
4. J. J. Hopfield, *Proc. Natl. Acad. Sci.* **79**, 2554 (1982).
5. P. W. Anderson, *Proc. Natl. Acad. Sci.* **80**, 3386 (1983).
6. F. E. Yates (editor), *Self-Organizing Systems: The Emergence of Order* (Plenum Press, New York, 1987).
7. D. Pines (editor), *Emerging Syntheses in Science* (Addison-Wesley, 1988).
8. D. Stein, *Proc. Natl. Acad. Sci.* **82**, 3670 (1985).
9. A. Georges, M. Mézard and J. S. Yedidia, *Phys. Rev. Lett.* **64**, 2937 (1990).
10. L. Onsager and S. Machlup, *Phys. Rev.* **91**, 1505, 1512 (1953).

The Physics of Neural Networks

Hanoch Gutfreund

*The Racah Institute of Physics, The Hebrew University of Jerusalem,
Jerusalem 91904, Israel*

and

Gerard Toulouse

*Laboratoire de Physique Statistique, Ecole Normale Supèrieure,
24 rue Lhomond, 75231 Paris Cedex 05, France*

1. Introduction

a. Historical Perspective

Neural networks are studied in order to understand intelligence, real and artificial. As for real intelligence, the aim is to discover how brains — living neuron systems — function. As for artificial intelligence, it is to invent new computing devices. That involves at least two disciplines: biology and computer science, and that means two prospects which may at times converge, at times diverge. This duality of interest has been present from the beginning of neural network science.

The study of neural network models has a history of five decades, if one agrees to mark its beginning with the work of McCulloch and Pitts[1] in 1943, who have introduced the notion of the formal neuron as a two-state threshold element and have shown that networks of such elements can implement any logical function. An important landmark in the early stages of this history is the work of Hebb,[2] who suggested that a concept is represented in the brain by the firing activity of a cell assembly and that learning proceeds by modification of the synaptic junctions connecting the neurons. This was followed by the study of a variety of models for associative memory, pattern recognition and various classification tasks. To name just a few of these developments, one should mention

7

the notion of the Perceptron — a feedforward network which can be trained to solve simple recognition problems,[3] the Adaline machine[4] — another such adaptive device, and the research programs of Amari,[5] Caianiello,[6] Grossberg,[7] Kohonen,[8] Kryukov,[9] and Palm.[10]

One recent novelty in the field has been the entry of physics as a third partner in between biology and computer science. This development may be traced back to the analogy between the activity of a neural network and the collective states of coupled magnetic dipoles made by Cragg and Temperley[11] in 1954, in a paper which did not get much attention. Twenty years later, Little[12] made the analogy between synaptic noise and temperature and suggested that persistent firing states of neural network dynamics appear just like the ordered phases in magnetic systems. He argued for the relevance and applicability of the concepts of statistical mechanics to the theory of neural networks. Hopfield[13] has completed the analogy with physics by introducing a concept of computational 'energy'. He studied networks with symmetric couplings, which are therefore described by a Hamiltonian, and showed the equivalence between the asymptotic dynamical behavior of such networks and equilibrium thermodynamical properties of random magnetic systems similar to spin glasses. The model suggested by Hopfield was subsequently solved analytically.[14] This was the first successful and non-trivial application of statistical mechanics in this field, giving rise to a variety of new and surprising results. The last development to be mentioned in this brief historical introduction is the work of Gardner,[15] who proposed a new approach, applying statistical mechanics in the space of possible networks. This allows the calculation of certain limits on the storage capacity and computational capability.

b. Why Statistical Physics?

Statistical physics achieved, during the last century, one of the most remarkable successes of science, viz., the explanation of thermodynamic laws in terms of statistical properties of large collections of atoms or molecules. More recently, it focused with renewed success on the emergence of collective behavior of large assemblies of elements reflected in the phenomena of phase transitions. Similar words can be used to formulate the basic question in neuroscience — "Mind from matter?" — as the emergence of collective 'intelligence' in a large assembly of neurons, each of which responds by well-defined rules to electrochemical inputs. Thus, the involvement of statistical physics seems quite natural. Moreover, neural networks process information, and the relation between information and thermodynamics has been realized long ago, ever since the work of von Neumann and Shannon.

Nevertheless, the application of the concepts and methods of statistical physics to neural networks is not straightforward. The interactions between neurons, transmitted through the synaptic junctions, are not symmetric. This is a basic difference from ordinary physical systems, in which the symmetry of the interactions and, hence, the existence of an energy function, which is a cornerstone in the formulation of statistical mechanics, is warranted by Newton's third law.

During recent decades, statistical physics has explored the behavior of heterogeneous systems, in contrast with the simpler case of homogeneous systems. In particular, much conceptual and technical progress has been made in the study of spin glasses, which are spin systems with random interactions. The road between neural networks and spin glasses has been paved by the recognition that a neuron, formalized as a threshold automation, is an element quite similar to a two-state (Ising) spin. A solid bridge between the two fields has been established in 1985, when the latest techniques of statistical physics were put to use in order to solve nontrivial neural network problems. Indeed, this bridge is a two way path, whereby statistical physics brings discoveries and surprises to neural network science, triggering further exploration, while in return it receives a lot, in terms of incentives and testing grounds. As a matter of fact, some spin glass models may turn out to have more relevance for neural nets than for real spin glass materials.

In brief, the study of neural nets by statistical physics has acquired a dynamics of its own. It will go on, even if the connection with neurobiological experiments remains loose for some time. However, despite this relative autonomy, physicists will be chronically looking for guidance into biology. The reason is simply that the space of models, which one can invent and study, is just too vast. Obviously, much of the excitement on neural nets, i.e., networks made of neurons interacting via synapses, comes from the fact that we know from biology that intelligent behavior can be obtained with collections of such elements. It is a safe bet to predict that many further important steps will be guided by hints of this kind.

c. Purpose and Outline of the Paper

The objective of the present paper is to put the entry of theoretical physics into neural network science within proper perspective, and accordingly:

1) explain what physics has contributed so far, describing some of its tools and motivations,

2) suggest that the participation of physics is more than a transient phenomena, and that it is likely to be a helpful factor in neural network science.

Indeed, this domain of science has suffered in the past from excess of high and low mood. In retrospect, it seems that some of the most damaging mishaps — the rise and fall of perceptrons, the hardware-software dispute, and their negative consequence — could have been avoided. So, besides the standard way that physics contributes to biology by improving the instrumentation, maybe the spirit of theoretical physics will help to bring the field of neural networks to scientific maturity.

One useful approach, specially in statistical physics, consists in the full exploration of simplified models. The topic of memory is a good example of a problem where considerable conceptual clarification has thus occurred, ahead of experimental investigations. But it should be reminded that these models are almost always abstractions remote from the reality of biological complexity, and that their main service is to raise new questions and to educate our investigations.

With these words of caution, we conceive this text as a complement to other attempts at interdisciplinary pedagogy, such as D. J. Amit,[16] V. Braitenberg,[17] P. S. Churchland[18] and P. Peretto.[19]

The selection of topics and original contributions, reviewed in this paper, is guided by the common motive of the collection of articles in this book. However, the word 'spin glasses', appearing in the book's title, should not be taken too literally, but rather as a code name for the transfer of methods from statistical mechanics of large systems with quenched disorder to the study of neural networks.

The first two sections, following the introduction, describe the basic ingredients and assumptions of the models, introduce the fundamental concepts and define the essential goals. Sections 4 and 5 review the two major triumphs of statistical mechanics in this field: The analytical treatment of the Hopfield model and related developments, and Gardner's approach, shifting the attention to the space of synaptic interactions, with its broad range of implications and applications. Sections 6 and 7 review some results on how neural network models account for two important aspects of learning: Categorization and generalization. Finally, in Sec. 8 we raise a few points concerning the relevance of the models to biology. We have excluded from this review models describing the storage and retrieval of temporal sequences of data (see, for example, Chap. 5 of Ref. 16).

2. Basic Elements of Neural Network Models

a. The Biological Neuron

The building blocks of neural networks are the neurons. In neural network models these neurons are reduced to simple logical elements with well-defined input-output relations. A description of the transition from the biological neuron to its model caricature requires a brief neurobiological introduction. We would like to state at the outset that this introduction will be restricted to the essentials. We realize that such a statement is itself not immune to criticism and objections. It biases the outcome by assuming knowledge of what are essential points in this extremely complex system, which should be adopted as guidelines in the construction of models. Nevertheless, we shall present what is believed to be the essential biological background without further apology.

The neurons are canonically divided into three parts — the dendritic tree which collects the input signals from other neurons, the soma which transforms this input into an output signal, and the axon which transmits the output signal to other neurons in the system. Such a typical neuron is shown schematically in Fig. 1. The neurons are connected by synapses which are the points of contact between branches of the axon of a presynaptic neuron and the dendritic tree or soma of a postsynaptic neuron. In the central nervous system of a higher mammal each neuron may receive about 10^4 synaptic inputs, and its axon forms about the same number of synaptic junctions with other neurons.

Neurons communicate via electrical signals — action potentials (or spikes) which propagate along the axons, preserving their shape and amplitude, until they reach the synaptic buttons where a chemical neurotransmitter is stored in a large number of vesicles. The action potential releases the contents of several such vesicles into the synaptic cleft. The molecules of the neuro-transmitter reach the membrane of the postsynaptic

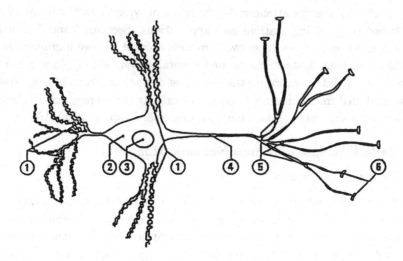

Fig. 1. Schematic representation of a typical neuron: 1 — dendritic tree; 2 — soma; 3 — nucleus; 4 — axon; 5 — axonal tree; 6 — synaptic buttons.

neuron where their reception induces a postsynaptic potential (PSP). The PSP's generated at the various points along the dendritic tree diffuse (with attenuation) towards the soma, where they are integrated. If the total sum of the PSP's, arriving within a short time interval, exceeds a certain threshold (of a few tens of millivolts) the neuron is likely to fire an action potential along the axon.

The contribution of a single presynaptic input to the PSP characterizes the synaptic efficacy. It is of the order of one millivolt, and can be either excitatory or inhibitory, namely, increasing or decreasing the likelihood of producing a spike. It should be emphasized that the PSP is not uniquely determined by the spikes in the input neuron. Several sources of noise related to fluctuations in the amount of the chemical transmitter released at the synaptic junctions result in probabilistic input-output relations.

We conclude this section with a few remarks on the time scales involved in neural dynamics. The typical time interval between the emission of a spike at the soma of the presynaptic neuron and the firing of the induced spike at the postsynaptic neuron is approximately 1–2 msec. After the emission of a spike the neuron cannot fire again during a refractory period of about 1–2 msec. This sets the maximal possible firing rate of 500–1000 spikes per second. In a network this maximal rate is reduced by a factor of

3–5. The reason is that immediately after the refractory period the neuron emerges with an abnormally high excitation threshold, significantly higher than the typical PSP's in the network. It is only when these threshold values decay to their normal level, after 7–10 msec, that the neuron can fire again. This is the relative refractory period. The maximal spike frequencies of 500–1000/sec are indeed observed in sensory neurons where the external stimulus can be arbitrarily strong. In a network, however, the typical firing rates are significantly lower, even in comparison with the maximal rates determined by the relative refractory period.

b. From the Biological to the Formal Neuron

The dynamics of the biological neuron is governed by the conservation of electric charge

$$C\frac{dU}{dt} = -F + I\,, \tag{1}$$

where U is the membrane potential, C is the capacitance, F is the membrane current and I is the sum of external inputs and synaptic currents from the other neurons in the network. The membrane current consists of several channels, each of which is characterized by a specific voltage-conductance relation and a specific time dependence. A realistic description of the neuron dynamics is very complicated. A simplified version leads to the classical Hodgkin-Huxley model, which has been one of the cornerstones of theoretical neurobiology.[20] Though simplified, it is still much too complicated to allow an analytical, or even numerical, treatment of large networks of Hodgkin-Huxley neurons. To make any progress in this direction, it is necessary to adopt a simpler description of the neurons. Several reduction procedures leading to simpler models, such as the integrate-and-fire models[20] or the FitzHugh-Nagumo model,[20] have been used by neurobiologists. However, the boldest reduction leading to the other extreme of neural models is the binary neuron, used in most studies of neural network models. The systematic reduction of the Hodgkin-Huxley neuron to the formal binary cell has been recently described by Abbott and Kepler.[21]

Comparing with the phenomenology of the Hodgkin-Huxley model, the binary neuron preserves the existence of the action potentials, but loses completely the description of the capacitive integration of the incoming current pulses and the recovery period following an action potential. It remains to be seen that what is left is sufficiently interesting and meaningful to justify the sacrifice in biological reality.

c. The Formal Neuron

In the model networks, to be discussed in this paper, the neurons, with their entire anatomical and biochemical complexity and diversity, are reduced to simple logical binary elements, which can be either in state $S = +1$ (active) or $S = -1$ (idle). The single neuron dynamics is determined by the value of the postsynaptic potential (PSP), accumulated on the membrane of neuron i at the end of a certain summation period:

$$U_i = \sum_j J_{ij}(S_j + 1) \tag{2}$$

where J_{ij} is the synaptic efficacy, measuring the contribution of a signal from the presynaptic neuron j. The elements of the synaptic matrix J_{ij} can be positive or negative, describing excitation or inhibition, respectively. The response of this neuron in the absence of noise is given by

$$S_i = \text{sign}(U_i - T_i) \tag{3}$$

where T_i is the threshold. We shall assume, for simplicity, that $\sum_j J_{ij} = T_i$, and define the local field at site i as

$$h_i = \sum_j J_{ij}S_j . \tag{4}$$

In reality, synaptic transmission is a noisy process and the PSP is not determined uniquely by the right-hand side of Eq. (2). The combined effect of several sources of synaptic noise leads[22] to a Gaussian distribution of the variable U_i

$$P(U_i) = \frac{1}{\sqrt{2\pi\delta^2}} \exp\left[\frac{U_i - \bar{U}_i}{2\delta^2}\right] \tag{5}$$

where the average $\bar{U}_i$ is the r.h.s. of Eq. (2) and the width δ is related to the parameters associated with the different factors of the synaptic noise. The probability that neuron i fires an action potential is equal to the probability that its PSP is higher than the threshold. This is approximated,[22] to a high accuracy, by the expression

$$P(S_i) = \frac{\exp(\beta h_i S_i)}{\exp(\beta h_i) + \exp(-\beta h_i)} \tag{6}$$

where β is proportional to the noise parameter δ. This expression is identical to the one which describes the single spin dynamics of an Ising system at temperature $T = \beta^{-1}$. When $\beta \to \infty$, which is the limit of zero noise ($T = 0$), this expression reduces to Eq. (3).

d. Network Architecture

Neural network models fall into two main classes of architecture, which represent two opposite limits:

a. A feedforward layered structure, named perceptron or multi-perceptron (according to the number of layers), in which the neurons are organized in successive layers without connections between neurons in the same layer, so that each layer receives inputs from the preceding layer only.

b. A self-coupled structure lacking any geometrical pattern in its connectivity. Here one can distinguish between the extreme case, with complete feedback, when each neuron is coupled to every other neuron and various degrees of diluted connectivity. The limit of sparse connectivity is obtained when the number of connections to each neutron does not scale linearly with the size of the network.

The mode of computation in the layered structure is very simple — it is performed by dynamic flux of activity from the first (input) to the last (output) layer. There is evidence that this is the dominant mode of computation in the early stages of visual processing (though there are many feedback connections whose function is not well understood to this day). Feedforward networks have been studied intensively for a variety of models and applications.[23]

In the self-coupled networks the input is supposed to clamp the initial state of the network, which subsequently evolves asymptotically under the network dynamics towards an attractor. The asymptotic behavior of the network represents the result of the computation. Such networks are named attractor neural networks, indicating that they perform computation via convergence into attractor states.

An aspect that is overlooked in many model structures is that a biological neuron transmits its information to a tiny fraction of the other neurons and that this spatial addressing of information may be selective enough to perform specific operations and transformations on the inputs. Horace Barlow (private communication) illustrates this point nicely with an analogy to lenses redirecting light rays into optical images. This important aspect is clearly beyond the reach of this paper.

e. Network Dynamics

We shall now concentrate on attractor networks. The state of the network is the configuration of activities at a given time, which in the case of binary neurons is expressed as an N-bit vector of the variables S_i. A state evolves in time along a trajectory in the space of 2^N possible configurations. This process depends on the rule, described above, which defines how a single neuron updates its present state, and also on the time relation between this event and the updating of the other neurons in the network. Two simplified versions of modes of updating, which represent two opposite limits, have been considered:

a) The first version assumes that all the neurons update their activity simultaneously at discrete time intervals t_n. The inputs of all the neurons at time t_n are determined by the activity configuration at time t_{n-1}.

b) The second version is the asynchronous dynamics in which the neurons are updated one after the other in a fixed or random sequence. In this mode an elementary time step of transition between two successive states is N^{-1} of the time step in the synchronous mode of updating. Asynchronous dynamics with the neurons being updated in a random sequence according to Eq. (6) is just the Glauber dynamics used in magnetic systems.

The network dynamics can be described by the $2^N \times 2^N$ transition matrix $W(I, J)$

$$W(I, J) = \frac{\exp\left(\beta \sum_i h_i^J S_i^I\right)}{\prod_i [\exp(\beta h_i^J) + \exp(-\beta h_i^J)]} \tag{7}$$

where h_i^J is the local field or pre-synaptic signal (Eq. (4)) reaching neuron i as determined by the activities in network state J, and S_i^I is the activity of this neuron in the consecutive state I. The transition matrix relates the probability $\rho(J, n)$ that the network be in state J at time t_n and the probability that it be in state I after one time step

$$\rho(I, n + 1) = \sum_J W(I, J)\rho(J, n). \tag{8}$$

The basic idea of attractor networks is that any cognitive response to an external stimulus, corresponding to an initial configuration of the network activity, is represented by a dynamical flow leading to an attractor of the network dynamics, corresponding to that stimulus. In general, the attractor could be a persistent state of activity or a more complex temporally varying state.

One of the main goals in the theory of such models is the study of the existence, abundance and nature of these attractors. Little[12] has studied the simpler case of persistent state attractors. He mapped the neural network, with synchronous dynamics, onto a two-dimensional Ising model in which each row contains the N spins and is labeled by the time step t_n. All the spins in one row are connected to the spins in the consecutive row by the interaction parameters J_{ij}. The appearance of an ordered phase in a two-dimensional Ising model is associated with the degeneracy of the largest eigenvalue of the transfer matrix, which is just the numerator of Eq. (6) with I and J denoting the spin configurations in successive rows. Little has shown that, likewise, the appearance of persistent states at large times implies the degeneracy of the largest eigenvalues of the transition matrix $W(I, J)$ and demonstrated numerically, in specific cases, the existence of a group of closely spaced eigenvalues of this matrix near unity, which are separated by a large gap from the other eigenvalues. However, it is not easy to relate these eigenvalues and their eigenvectors to the properties of the corresponding attractors, and the study of attractors in neural network models took a different direction, emphasizing explicitly the network states themselves.

3. Basic Functions of Neural Network Models

Model neural networks have been used to demonstrate possible scenarios for cognitive functions like memory and learning, to perform a variety of recognition, classification and optimization tasks, and to model neurobiological and psychophysical phenomena. In the present paper we shall review some of the main applications of neural network models, all of which have one common feature — the existence of a suitably defined 'energy' or cost function. These are the topics where statistical mechanics, so far, made its major contributions. To set the scene for a more detailed discussion, we begin with a brief treatment of the basic themes.

a. Associative Memory

It is assumed that memorized concepts are represented by specific configurations of the network activity, to be referred to as memories or stored patterns. Thus, a concept μ is represented by the N-bit vector $\{\xi_i^\mu = \pm 1\}(i = 1, \ldots, N)$. The network may serve as a model of associative memory if a set $(\mu = 1, \ldots, p)$ of such configurations are attractors of the network dynamics. This means that, starting from an initial configuration $\{S_i\}$, which has a sufficiently large overlap with one of the stored patterns, the system will flow to a fixed point of the dynamics, which is either the pattern itself or a configuration with a high overlap with that pattern. In the presence of noise the system will fluctuate in a close neighborhood of that pattern in configuration space. The time evolution of the system can be characterized by the overlaps of the network states, at time t, with the stored patterns

$$m_\mu(t) = \frac{1}{N} \sum_i \xi_i^\mu S_i(t) . \tag{9}$$

An external stimulus is recognized as the concept μ if it imposes an initial state, which after some time drives the network to the attractor corresponding to this concept, namely, if asymptotically $m_\mu \approx 1$ and all the other overlaps are small. The smallest value of $m_\mu(t = 0)$, which still leads to the corresponding attractor is a measure of the radius of the basin of attraction.

b. Learning

Learning in the context of layered neural network models means, in general, a procedure for finding synaptic connections J_{ij} between neurons i and j, belonging to two successive layers, which lead to the prescribed input-output relations. In self-coupled networks the goal of learning is to find a matrix J_{ij}, which ensures a dynamical behavior leading to specific attractors. The task is to organize the space of network states in basins of attraction around *a priori* known memory states. Note that this problem is the inverse problem to the general goal in the physics of collective phenomena, which is to find the equilibrium or asymptotic behavior when the interactions are known.

There are two lines of approach to this problem. One is to assume a specific dependence of the J_{ij}'s on the stored patterns. The simplest choice, in a self-coupled network is

$$J_{ij} = \frac{1}{N} \sum_{\mu=1}^p \xi_i^\mu \xi_j^\mu . \tag{10}$$

This form can be interpreted as a naive realization of Hebb's hypothesis that learning proceeds by the modification of the synaptic efficacies, determined by a correlation of the pre- and post-synaptic activities. Learning a concept, represented by an activity state $\{S_i\} \approx \{\xi_i^\mu\}$ means that the network is repeatedly driven by a strong external stimulus into this state and the synaptic coupling J_{ij} is thereby changed, according to Hebb's hypothesis, by

$$\Delta J_{ij} = \langle S_i \rangle \langle S_j \rangle = c \xi_i^\mu \xi_j^\mu \,. \tag{11}$$

where $\langle \cdots \rangle$ denotes a time average and c is a proportionality constant. Starting from a 'tabula rasa' and learning p such concepts, one gets Eq. (10) (with the particular choice $c = 1/N$). This form of the synaptic matrix is the basis of Hopfield's approach[13] to be discussed in Sec. 4. Adopting a specific *a priori* dependence of the synaptic matrix on the learned patterns is not really a learning process and may better be described as a storage prescription.

Intuitively, learning corresponds rather to a procedure which treats the J_{ij}'s as variables and modifies them by a learning algorithm to satisfy certain constraints. In particular, constraints which guarantee that a set of given memories become fixed points of the dynamics (at zero noise) are

$$\Delta_i^\mu \equiv \xi_i^\mu \sum_j \frac{J_{ij}}{\sqrt{N}} \xi_j^\mu > \kappa \tag{12}$$

for $i = 1, \ldots, N$; $\mu = 1, \ldots, p$, subject to the condition

$$\sum_j J_{ij}^2 = N \,. \tag{13}$$

The parameters Δ_i^μ are proportional to the absolute values of the local fields on site i when the network is in state $\{S_i\} = \{\xi_i^\mu\}$. The configurations $\{\xi_i^\mu\}$ are fixed points of the dynamical Eqs. (3) even for $\kappa = 0$, but a finite κ is needed to ensure significant basins of attraction. The normalization is imposed to avoid the freedom in the value of κ due to an overall scaling of Eq. (12). One specific algorithm for finding a matrix J_{ij}, which satisfies Eqs. (12) and (13), is the perceptron algorithm, originally formulated for a perceptron network.[3] It proceeds as follows. Starting from an arbitrary matrix J^0, the Np inequalities (12) are examined one after the other and whenever one of them is not satisfied, all the J_{ij}'s, for the corresponding site i, are modified to

$$J_{ij} \rightarrow J_{ij} + \frac{\lambda}{N}\xi_i^\mu \xi_j^\mu \tag{14}$$

where λ is a positive step-size parameter. Following each updating, the new J_{ij}'s are normalized to preserve Eq. (13). It was shown by Gardner[15] that this algorithm can be applied to multiply connected networks and that the convergence theorem, which guarantees convergence to a solution if such a solution exists, can be extended to this case.

An important contribution of Gardner was to show that the question of the existence of solutions to the learning task, defined above, and thus the problem of the storage capacity, can be decoupled from the problem of finding such a solution by a specific learning algorithm. This development and its consequences will be reviewed in Sec. 5.

c. Categorization

Categorization is a particular aspect of learning leading to an efficient organization of the learned data. It is the ability to form class-concepts from a number of individual entities sharing common features. In the context of attractor neural networks categorization will take the following sense: Suppose that a network has learnt a sufficiently large number of patterns $\{\xi_j^\mu\}$, all of which have a significant overlap with a particular pattern $\{\xi_i^0\}$. The patterns $\{\xi^\mu\}$ can be considered to be variants of the prototype $\{\xi^0\}$. The question is whether the network will also form an attractor corresponding to the prototype configuration to which it has not been exposed directly (prototype formation). This question was addressed by Virasoro,[24] who showed that this is indeed the case for a typical learning procedure. Moreover, the prototype is stored with larger stability than its explicitly learnt variants (prototype dominance). This is a remarkable property of self-organization of a neural network memory. The issue of categorization is discussed in Sec. 6.

d. Generalization

Generalization is an integral aspect of learning. It is the ability to extract a rule from examples. Generalization has been a central theme in the study of feedforward networks. A typical classification problem is to design a network which assigns the correct output to every possible pattern in a very large input space. It is inconceivable to perform an exhaustive training of the network in the entire input space. What is done instead is to adjust the weights by learning independently drawn examples of input-output pairs belonging to a training set of limited size. This training process, known as supervised learning, is aimed at minimizing an error measure over the training set, and can be analyzed by Gardner's method. The question then is: What is the proportion of correct responses to inputs outside of the training set? This defines the concept of generalization error, which is the counterpart of the training error on the new inputs. For many years there was no theoretical framework to study this question and the answer depended very much on the specific architecture of the network (number of layers and number of neurons in each layer) and on the specific problem.

Recently, methods of statistical mechanics have been applied to the problem of generalization and the first results have emerged. In Sec. 7 we shall review the basic results on generalization by a perceptron.

e. Optimization

The contributions of statistical mechanics to the study of hard optimization problems are outside of the scope of the present review. Nevertheless, we mention this topic briefly, confining our remarks to the present section only, in order to point out an appealing example of computation with attractors.

Optimization problems are characterized by a cost-function depending on a set of relevant variables, and the task is to find a configuration of these variables for which the cost-function attains its global minimum. Perhaps, the best known of these problems is the Traveling Salesman Problem (TSP), which involves a set of N points with given distances between any two of them. The task is to find a tour of minimum length passing through all the points only once. In this example the cost-function is the length of the trajectory and the variables are the $N!$ permutations of the given points. This problem belongs to a class of NP-complete problems, which means that no algorithm is known which is guaranteed to find a solution in a time increasing with N slower than exponentially. The complexity of the problem stems from the large number of local minimal, implying that any steepest descent procedure to minimize the cost-function is likely to terminate in one of them.

The analogy between the cost-function with its many local minima and the free-energy landscape in a disordered system, notably the spin-glass, has attracted the interest of physicists to optimization problems. On closer inspection the similarities are even greater. Not only is there a multitude of local minima in the TSP and the spin-glass, but the landscape has in both cases, qualitatively, the same ultrametric structure.[25] This suggests that methods of statistical mechanics for computing the ground state energy of a spin-glass can be applied to estimate typical values of the lowest 'energy' in several optimization problems.[26]

Comparison with relaxation processes in disordered physical systems suggests that an analogue of the process of annealing, to reach an equilibrium in a system with many metastable states, may be used to find the minimum of the cost-function of an optimization problem. This led to the method of simulated annealing,[27] which introduces noise into a dynamical procedure of reducing the cost-function. One defines a fictitious temperature $T = \beta^{-1}$ and performs consecutive changes of the relevant parameters. A change ΔJ is accepted with the probability

$$P = [1 + \exp(\beta \Delta E)]^{-1} \tag{15}$$

where ΔE is the change in the cost-function induced by ΔJ. This also allows steps, which (occasionally) increase the cost-function, and makes it possible to escape from local minima. In order to reach an optimal solution it is necessary to reduce gradually the 'temperature' T. The actual 'cooling' prescription requires a careful analysis, but significant improvements over existing methods of optimization have been achieved in a variety of applications.

More recently, it has been proposed that the relaxational dynamics in an attractor neural network can be used to solve optimization problems.[28] To this end one has to design a network in which every neuron is identified in terms of the variables of the problem. For example, in the TSP each neuron represents one of the points and a position in a tour (thus, N^2 neurons are needed in this case). The allowed states of the network are configurations in which N neurons are active, so that each given point and every possible position appear only once (the constraints of the problem). The parameters of the problem (distances between points) and the constraints are included in the synaptic efficacies. The dynamics of the network is described by an 'energy', which has global minima at the activity configurations corresponding to the optimal trajectories. The hope is that the 'energy'–landscape of the representative network has none or significantly less local minima (spurious states) than the cost-function in the original problem. In that case, an initial state, corresponding to a first guess, will evolve under the network dynamics to a desired solution.

The actual scheme, proposed in Ref. 28 for the TSP, performs reasonably well for small values of $N(N = 10-30)$ and fails badly for large N, indicating that there are still many spurious states in this network. Perhaps the addition of an annealing procedure to the network dynamics may improve the performance of this model. However, what is important is not the failure to provide a new algorithm for the TSP, but the possibility of optimization by attractor dynamics. After all, biological neural networks do not seem to be very efficient in solving the TSP, or other NP-problems. On the other hand, certain tasks in cortical computation (such as binocular matching) may be formulated as optimization problems and 'solved' spontaneously by specific networks, in which the solutions are represented by attractors of the dynamics.

4. The Hopfield Model

The basic element in Hopfield's model is the assumption of a symmetric synaptic matrix. In general, when $J_{ij} = J_{ji}$, the asynchronous stochastic dynamics of the network is governed by an energy function of the form of an Ising spin system.

$$E = -\frac{1}{2} \sum J_{ij} S_i S_j \,. \tag{16}$$

The existence of an energy implies that, at $T = 0$, the system flows in configuration space to local minima of E. At $T \neq 0$, the system relaxes to an equilibrium described by the Gibbs distribution of states $\exp(-\beta E)$, and the attractors of the dynamics are minima of the free energy. Thus the study of the asymptotic dynamics is reduced to a problem in equilibrium statistical mechanics.

The specific form of J_{ij} in Hopfield's model is given in Eq. (7), with the provision that the stored patterns $\{\xi_i^\mu\}$ are random, uncorrelated and unbiased (equal probability of $+1$ and -1). It can be shown, by a simple signal to noise analysis,[28] that as long as their number (p) does not exceed $N/(2\ln N)$, they are fixed points of the dynamics at $T = 0$ or global minima of the energy, surrounded by very large basins of attraction. As p increases, the random correlations between the memories create noise which acts to destabilize them as minima and to reduce their basins of attraction. The question is what happens when p becomes proportional to N, $p = \alpha N$. This question was answered by Amit *et al.*,[14] who showed that the free energy of the Hopfield model can be calculated and analyzed using methods, concepts and analogies of the statistical mechanics of disordered systems.

a. Solution of the Model

One is interested in typical results averaged over the possible realizations of the stored patterns. The randomness of the stored patterns and the extensive connectivity of the network establish an analogy with physical systems that possess the ingredients of quenched disorder and long-range interactions, in particular, with the Sherrington-Kirkpatrick (SK) model[30] of a spin glass. The average over quenched disorder is usually performed by the replica method[31] and the long-range interactions imply the validity of the mean field theory.

The different modes of behavior are characterized by the order parameters

$$m_\mu = \frac{1}{N} \sum_i \xi_i^\mu \langle S_i \rangle \tag{17}$$

and

$$q_{\alpha\beta} = \frac{1}{N} \sum_i \langle S_i^\alpha \rangle \langle S_i^\beta \rangle \tag{18}$$

where $\langle \ldots \rangle$ is the thermal average. The parameters m_μ are the equilibrium values of the overlaps defined in Eq. (8) and $q_{\alpha\beta}$ is the overlap between two replicas. In the replica symmetric approximation it reduces to a single parameter, q, which is the equivalent of the Edwards-Anderson order parameter in a spin-glass.[32] The overlap parameters m_μ are the new features (compared to the SK-model). Memory retrieval is possible only in phases where $m_\mu \neq 0$. Let us summarize the main results (the numbers quoted in a and b are obtained with one step replica symmetry breaking)[33]:

a) At $T = 0$ and at a finite storage level α, the memories are not fixed points, but new fixed points appear in their close neighborhood, as long as α is below the critical value $\alpha_c = 0.144$.

b) The overlap of these fixed points with the memories determines the fraction of retrieval errors $f = 1/2(1 - m_\mu)$. At α_c, one gets $f \approx 0.01$. The percentage of errors increases with α and jumps discontinuously, at α_c, to 50%, corresponding to a zero overlap with the learned patterns, indicating a complete loss of memory (Fig. 2b). This 'forgetting catastrophe' is clearly undesirable in a realistic memory model.

c) The phase diagram in the $(\alpha - T)$-plane is shown in Fig. 2a. As T increases, at a fixed value of α, the overlap with the memories decreases until it drops discontinuously to zero at T_M. The retrieval phase below the line T_M is characterized by valleys with macroscopic energy barriers surrounding the memories. It coexists with the spin-glass phase ($q \neq 0$), characterized by exponentially many local minima. The spin-glass phase persists up to the line T_G.

d) The attractors, corresponding to the memories, become global energy minima only below the line T_C. This is the line at which a true thermodynamic phase transition takes place.

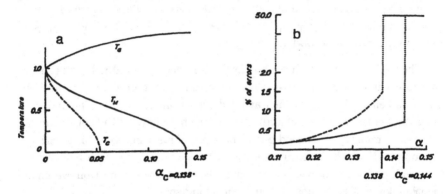

Fig. 2. a) Phase diagram of the Hopfield model. See text for the meaning of the different lines. b) Percentage errors in retrieval. Broken line: The replica symmetric result; full line: effects of one-step replica symmetry breaking.

b. The Merit of the Hopfield Model

In order to apply the formalism of statistical mechanics, it was necessary to assume synaptic symmetry, which is biologically a very unrealistic hypothesis. The channels of communication in real neural networks are unidirectional and there is no reason that neuron i will be affected by neuron j as neuron j by neuron i. In fact, the very existence of a connection from neuron i to neuron j is not implied by the existence of a connection in the opposite direction. Thus, the assumption of synaptic symmetry is a step backward from the point of view of biological plausibility and was frequently criticized by neurobiologists, casting doubt on the entire approach. In retrospect, the Hopfield model turned out to be profitable both for statistical physics and for neutral network theory.

From the point of view of statistical mechanics, the Hopfield model has several interesting features:

a) It is another solvable and rich model of a random long-range system.

b) It spans a whole range of intermediate modes of behavior, between the infinite range Ising ferromagnet when $p = 1$ (the Mattis model[34]) and the SK spin-glass[30] when $\alpha \to \infty$.

c) It provides an instructive example where one can go beyond the replica symmetric approximation. It turns out that single-step replica symmetry breaking[33] gives corrections to the replica symmetric results, in very good agreement with numerical simulations.

d) The existence of macroscopic free-energy valleys which are not ground states of the system, at least between the lines T_C and T_M, is a novel property of this model.

The major contribution of the Hopfield model to neural network theory was to open a large galley of concepts, techniques and analogies, and to direct the effort in a way which avoided the initial obstacles. The analysis of the model leads to results and insights which go beyond the constraint of synaptic symmetry. It turned out to be a useful starting point for a variety of modifications which removed some of the constraints and drawbacks of the original formulation of the model. In the following section we shall mention some of these post-Hopfield developments.

The Hopfield model is inherently connected with the self-coupled network architecture. It is worth mentioning that a layered-network model, with synaptic couplings between successive layers given by Eq. (10), can be analyzed by similar methods of Statistical Mechanics.[35]

c. Beyond the Standard Model

We refer to the model described above, with the specific choice of the J_{ij}'s given by Eq. (10), as the standard model. We have already mentioned several drawbacks of this form of the synaptic matrix — it is symmetric, it is valid only for uncorrelated patterns and it leads to the 'forgetting catastrophe'. We wish to point out that none of these difficulties occurs in the Gardner approach, to be discussed in the next chapter. However, it is instructive to show that they can also be overcome by simple modifications of the standard model.

1. *Robustness to random asymmetry.* It has already been pointed out by Hopfield that the addition of random asymmetry to a symmetric matrix J_{ij} is equivalent to a new source of stochastic noise. To demonstrate this, assume that at some point during the evolution of the network, the variable S_k is not aligned with its local field, namely,

$$S_k \sum_j J_{kj} S_j < 0 \,. \tag{19}$$

This implies that S_k will now be changed to $-S_k$ and the corresponding change in the 'energy' is

$$\Delta E = \sum_j J_{kj} S_j S_k + \sum_j J_{jk} S_j S_k \,. \tag{20}$$

The first term is always negative, in view of Eq. (19). The second term is equal to the first when $J_{ij} = J_{ji}$ and E never increases. When $J_{ij} \neq J_{ji}$, the system sometimes makes transitions, which increase the energy. This is just what happens in the presence of noise. What is then the effect of this noise on the stability of the attractors corresponding to the memories? This question was first studied by numerical simulations in the context of the Hopfield model with broken bonds.[36] Starting from a fully connected network, in which $p = \alpha N$ memories have been stored using the Hebb rule (Eq. (4)), one chooses at random a fraction of pairs (i, j) and sets either J_{ij} or J_{ji} to zero. This introduces a random asymmetry of the matrix J_{ij}. It turns out that the storage capacity and the size of the basins of attraction are somewhat reduced but, on the whole, the performance of the network as an associative memory is quite robust to this operation. For example, when all the bonds are made unidirectional, the critical value of α is reduced from 0.14 to 0.076. Thus, the performance of the network as an associative memory is quite robust to the noise introduced by random asymmetry.

The role of random asymmetry was also emphasized, in a different context, by Parisi.[37] In a symmetrical network there are, outside of the basins of attraction of the memories, infinitely many fixed points representing the spin-glass phase. Parisi suggested a distinction between a 'meaningful' operation of the network, when, in response to a recognizable external stimulus, the system converges to one of the memories, and a 'meaningless' behavior, when the system wanders around chaotically in configuration space.

This raises the question of the effect of asymmetry on the spin-glass phase. Parisi conjectured and demonstrated by simulations that asymmetry of the synaptic matrix suppresses the spin-glass phase. The suppression of the spin-glass phase was proved analytically in the case of extreme asymmetry, when J_{ij} and J_{ji} are completely uncorrelated,[38] and was suggested to occur at any finite degree of asymmetry.[39] This motivated a detailed study of the dynamics of asymmetric spin-glasses and of the asymmetric Hopfield model[40] which generally supports this conclusion. How this occurs at small asymmetries remains, however, a subtle question.[41]

2. *Correlated memories.* The simplest form of correlation between the patterns appears when the ξ_i^μ's chosen to be $+1$ or -1 with probability

$$P(\xi = \pm 1) = \frac{1}{2} \left(1 \pm m \right). \tag{21}$$

In this case the pattern has a finite bias

$$m = \frac{1}{N} \sum_i \xi_i^\mu \tag{22}$$

and any two patterns are correlated with each other as reflected by the finite overlap between them

$$\langle \xi_i^\mu \xi_i^\nu \rangle \equiv \frac{1}{N} \sum_i \xi_i^\mu \xi_i^\mu = m^2 \quad (\mu \neq \nu). \tag{23}$$

The biological motivation to study biased patterns is the experimental evidence that the mean level of spatial activity in the cortical areas, in which associative functions are detected, is relatively low. In Sec. 8 we shall discuss some experimental findings on cortical activity and their implication for the notion of computing with attractors. At present, we mention the low level of spatial activity only as a compelling reason for being able to treat patterns which are strongly biased to -1 (actually, in the standard model there is a $(+1, -1)$-symmetry resulting in an equivalence between high $(m \leq +1)$ and low $(m \geq -1)$ activity, but this is an artifact due to oversimplification of the formulation). An additional motivation to study biased patterns is to make contact with the extensive work on sparse coding in a variety of hetero- and auto-associative networks.[42-45]

The failure of the standard model to accommodate biased patterns is due to the fact that in the form of J_{ij} in Eq. (10), every neuron is affected by a finite average noise contributed by each pattern. The result is that, for example, for $m = 0.5$ only $p = 5$ memories can be stored in the network even as $N \to \infty$. A simple modification of Eq. (10), which eliminates this average noise, is

$$J_{ij} = \frac{1}{N} \sum_{\mu} (\xi_i^\mu - m)(\xi_j^\mu - m). \tag{24}$$

This form implies that when the system is in the state defined by the μ-pattern, the local field on site i has a signal-term ξ_i^μ and a Gaussian noise-term with zero mean and a width which increases with the number of stored patterns p. This guarantees that, if p is not too large, then ξ_i^μ is the stationary state of neuron i. Without the subtraction of the bias m, the noise-term has a finite average, which (at certain sites) reduces significantly the signal term. When Eq. (24) is complemented by a constraint, restricting the dynamics of the network to a subspace of configurations with the same activity as the attractors, one gets a model for the storage and retrieval of biased patterns, which can be solved and analyzed in great detail, just like the standard model.[46] The storage capacity increases with m, up to $m \approx 0.925$, and then drops rapidly to zero when $m \to 1$.

There exists a much more effective scheme for biased patterns in the sparse-coding limit, which results in a diverging storage capacity when $m \to 1$, instead of the vanishing one found above. This limit is treated more naturally in the binary representation $(0, 1)$. The latter is related, up to a term $\sum_j J_{ij}$, by a simple transformation to the $(+1, -1)$-representation used here. We have assumed, in Sec. 2c, that this term is cancelled by a site-dependent threshold T_i. If, instead, one properly adjusts a uniform threshold, one finds that the critical storage capacity diverges as $m \to \pm 1$,

$$\alpha_c = -\frac{0.5}{(1 - |m|)\ln(1 - |m|)}. \tag{25}$$

It is outside of the scope of the present paper to discuss the relation of this result with other sparse-coding models. We mention it here in anticipation of a similar result to be obtained in the Gardner approach.

More complicated correlations arise when the patterns are grouped in several 'similarity' classes, which themselves may include several subclasses. Such structures are characterized by hierarchical correlations. Again, there is ample motivation to study such correlations, and several proposals to treat them by specific modifications of the standard model have been proposed.[47-50] We shall, however, defer the discussion of hierarchical correlations to Sec. 6.

In concluding this subsection on correlated patterns, we wish to point out that there exits, within the general approach of an *a priori* prescribed synaptic matrix, a model which can store a set of patterns with any type of correlations. This is the projection (also called pseudoinverse) model[51]

$$J_{ij} = \frac{1}{N} \sum_{\mu, \nu} A_{\mu\nu}^{-1} \xi_i^\mu \xi_j^\nu \qquad (26)$$

where $A_{\mu\nu}$ is the matrix of correlations between the patterns

$$A_{\mu\nu} = \frac{1}{N} \sum_i \xi_i^\mu \xi_i^\nu . \qquad (27)$$

This model has been studied analytically (with the additional assumption of vanishing self-coupling, $J_{ii} = 0$) by the same methods as the Hopfield model.[52] At first glance this form of synaptic efficacies has the biological disadvantage of being nonlocal, however, there exists a local learning algorithm, which leads exactly to this synaptic matrix.[53] A disadvantage of the pseudoinverse rule, from the point of view of psychophysical relevance, is that it does not have the generic property of categorization, which was mentioned in Sec. 3d and will be discussed in Sec. 6.

3. *Avoiding the 'forgetting catastrophe'.* The reason for the sudden destruction of the entire network memory at a critical storage level is the equivalence of the learnt patterns in the synaptic efficacies given by Eq. (10). Since all the patterns are stored with equal strengths they are also destroyed simultaneously, when the noise due to excessive storage is sufficiently large to destabilize a memory attractor. To provide a mechanism for selective 'forgetting' of memories it is necessary to design networks in which the patterns are stored with unequal intensities. Such models have the properties of 'palimsests', namely, recently acquired memories erase and replace the old ones.[54]

One way to achieve this goal is to adopt a strategy of 'marginal learning'. The basic idea is to add a new pattern to the network by modifying the existing synaptic efficacies, according to the Hebb rule, but with a weight which is just sufficient to overcome the destabilizing effect of the noise produced by the previously learnt patterns. The noise in the network is characterized by the width of the distribution of the synaptic efficacies.

$$K = \langle J_{ij}^2 \rangle - \langle J_{ij} \rangle^2 . \qquad (28)$$

The storage intensity of the new pattern is measured by

$$k = \langle \Delta J_{ij}^2 \rangle - \langle \Delta J_{ij} \rangle^2 \,. \tag{29}$$

It turns out, mainly from numerical simulations, that the last pattern is retrievable, with an error not exceeding 1.5%, if

$$\frac{k}{K} = \frac{\varepsilon^2}{N} \tag{30}$$

with $\varepsilon \approx 2.5$. With this value of ε it is possible to add new patterns to the network indefinitely, 'remembering' only a fraction of them, the ones acquired most recently.

A model, which shows this behavior and can be treated analytically,[55] has synaptic couplings of the form

$$J_{ij} = \frac{1}{N} \sum_{\mu=1}^{p} \Lambda \left(\frac{\mu}{N} \right) \xi_i^{\mu} \xi_j^{\mu} \tag{31}$$

with

$$\Lambda(x) = \varepsilon \exp \left(-\frac{x\varepsilon^2}{2} \right) \,. \tag{32}$$

The last pattern corresponds to $\mu = 1$ and the first one to $\mu = p$. The patterns are stored with exponentially increasing (from first to last) intensity. The storage capacity and the retrieval error depend on the value of ε. An optimal value is found around $\varepsilon_{opt} \approx 4.1$, for which about $0.05N$ of the patterns, the ones which have been learnt most recently, are retrieved with less than 1.5% error.

Another way of avoiding the 'forgetting catastrophe' is by imposing bounds on the size of the synaptic efficacies.[54,56] Memories are added to the network one after the other and the J_{ij}'s are modified according to the standard model as long as they have not reached the maximal allowed value. The value of J_{ij} upon adding the pth pattern is

$$
\begin{aligned}
J_{ij}(p) &= J_{ij}(p-1) + \frac{\lambda}{N} \xi_i^p \xi_j^p \quad &\text{for } |J_{ij}(p)| < A \\
&= J_{ij}(p-1) &\text{otherwise}\,.
\end{aligned} \tag{33}
$$

This model has the basic features of a 'palimpsest' memory, provided that the ratio between the storage intensity λ and the efficacy bound A is neither too small (then, the model reduces to the standard model) nor too big (then, only the last pattern is 'remembered').

The simplicity of these models for a memory that learns and forgets suggests considering their possible psychobiological relevance. One of the basic empirical findings on the short term memory of humans is that they can recall sequences of 7 ± 2 items shortly after their presentation. It is tempting to relate this figure to the number of retrievable patterns, predicted by these models to be fN, where $f \approx 0.01$–0.05 and N should be interpreted here (actually, in all the results on storage capacity) as the number of connections to a neuron. Although the actual coefficient should not be taken too seriously, these models suggest a relation between the short term memory capacity and the connectivity.[57]

5. The Gardner Approach

a. A Microcanonical Formulation

Let us now turn to the basic problem defined in Sec. 3b, that of the existence of a synaptic matrix J_{ij} which satisfies, for a given set of patterns, the inequalities (12), subject to condition (13). Since the J_{ij}'s are continuous variables, this problem can be reduced to finding the volume in J-space for which these conditions are satisfied. This volume, expressed as a fraction of the total volume of properly normalized synaptic matrices, can be written as

$$V = \frac{\prod_i \int d\Omega_i \delta(J_i^2 - N) \prod_\mu \theta(\Delta_i^\mu - \kappa)}{\prod_i \int d\Omega_i \delta(J_i^2 - N)} \qquad (34)$$

where $d\Omega_i = \Pi_j dJ_{ij}$, J_i is the ith row of the synaptic matrix, the parameters Δ_i^μ have been defined in Eq. (12), and $\theta(x)$ is the step function: $\theta(x) = 1$ when $x > 0$, and $\theta(x) = 0$ otherwise.

In the absence of any restriction on the correlation between J_{ij} and J_{ji}, the sites i are decoupled and the fractional volume can be calculated for each site separately. As usual in statistical mechanics, we are interested in the typical case among all the possible representations of the stored patterns and we should, thus, average $\ln V$ over the probability distribution of ξ_i^μ. This distribution reflects any possible structure of correlations within the set of patterns. In the case of uncorrelated patterns the ξ_i^μ's are chosen independently to be equal to $+1$ or -1, with equal probability. The average over $\ln V$ requires again the use of the replica method. The basic parameter which appears in the calculation is the overlap between solutions in two different replicas

$$q_{\alpha\beta} = \frac{1}{N} \sum_{ij} J_{ij}^\alpha J_{ij}^\beta. \tag{35}$$

In the replica symmetric approximation this reduces to a parameter q which measures the average overlap between two possible solutions. As the number of memories increases the volume of solutions decreases and q increases until, when the volume shrinks to zero, only one solution remains (since, assuming replica symmetry, the volume is convex) and $q = 1$. This limit defines the critical storage capacity α_c. For uncorrelated pattern sat $\kappa = 0$, one finds $\alpha_c = 2$. The replica symmetric solution is stable at $\alpha_c(\kappa), \kappa > 0$; it becomes marginally stable for $\kappa = 0$, and unstable for $\kappa < 0$. The curve $\alpha_c(\kappa)$ is shown in Fig. 3.

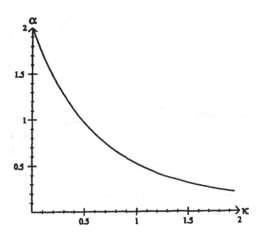

Fig. 3. Critical storage capacity as function of κ.

The decoupling of the sites i implies that the results of the calculation are valid for each site, separately. Thus this result provides, at the same time, the storage capacity of a perceptron with N (as $N \to \infty$) input units and a single output unit. This problem was studied previously by a combinatorial-geometrical method and the result $\alpha_c(\kappa = 0) = 2$ has been known for a long time.[58,59] The method of statistical mechanics seems to be more powerful as it allows us to obtain results for any value of κ and also for a variety of constraints on the synaptic parameters and on the structure of the patterns. It is not clear how (and if) the other method can be generalized to treat all these cases.

b. The Case of Biased Patterns

Gardner has also calculated the storage capacity for biased patterns, which satisfy Eq. (21). The storage capacity increases with m and diverges, when $m \rightarrow \pm 1$, like (Fig. 4)

$$\alpha_c = -[(1 - |m|) \ln (1 - |m|)]^{-1} \tag{36}$$

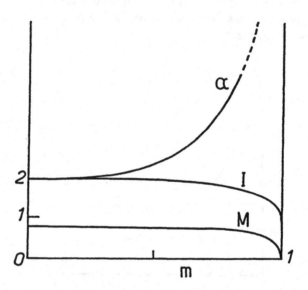

Fig. 4. Critical capacity, typical value of the parameter M and information content I as functions of the bias m for $\kappa=0$.

which is larger by a factor of 2 than the result quoted in Sec. 4c (Eq. (25)). This diverging storage capacity should not be taken as a miracle, since the information content per pattern decreases in the same proportion, with the overall result that the total information stored, actually, decreases slightly with m. In the course of this calculation a new order parameter appears, which measures the bias in the synaptic weights,

$$M = \frac{1}{N} \sum_j J_{ij} \, . \tag{37}$$

Typically, $M > 0$ for $m > 0$, which means that the pattern $\{1, 1, \dots, 1\}$ is stable. We mention this point in anticipation of the discussion on categorization in Sec. 6.

c. A Canonical Formulation

It is possible, and usually more convenient, to cast the problem of existence of solutions to the inequalities (12) into a form which resembles the canonical ensemble treatment in statistical physics. This was first done by Gardner and Derrida.[61] To this end, one views the learning process as the minimization of a properly defined 'energy', per site, which can generally be written as

$$E = \sum_\mu e(\Delta^\mu) \tag{38}$$

where we have suppressed the site index i. One possible definition of the function E is the sum of unsatisfied constraints (12). In this case

$$e = \theta(\kappa - \Delta^\mu) \tag{39}$$

and the 'energy' is simply the number of errors in learning. Other choices of the 'energy' will be mentioned in Sec. 5d.

It is assumed that the system is in contact with a 'heat-bath' at 'temperature' $T = \beta^{-1}$, so that each configuration of the J_{ij}'s appears with a Gibbs probability

$$P(\{J_{ij}\}) = \frac{1}{Z} \delta(J_i^2 - N) \exp(-\beta E\{J_{ij}\}) \tag{40}$$

where Z is the partition function, namely, the numerator integrated over the space of J_{ij}'s. The typical ground-state energy, averaged over the space of possible patterns, is given by

$$E_0 = - \lim_{\beta \to \infty} \frac{d}{d\beta} \langle \ln Z \rangle_\xi \, . \tag{41}$$

The averaging of $\ln Z$ over ξ is performed, again, by using the replica method. When α is small enough, one finds that $E_0 = 0$. The critical α is obtained at the point where this 'energy' becomes finite.

d. Constraints on the Synaptic Weights

The formalism presented in the last two sections can be extended to models with local constraints on the individual J_{ij}'s rather than the global constraint of normalization, represented by the δ-function in Eqs. (34) and (40). One class of models of this nature is distinguished by J_{ij}'s which are only allowed to assume a discrete set of values. In other models the J_{ij}'s may be continuous, but restricted to intervals with specific bounds, or may be chosen with an *a priori* given probability distribution.

We shall confine our discussion to models with discrete couplings. The study of such models is well motivated, both from the biological and the practical point of view. It is implausible to assume a biological mechanism which preserves the infinite precision of truly continuous J_{ij} and it is therefore interesting to study the effect of some coarse graining by encoding the information using a finite number of discrete values of J_{ij}. Likewise, in hardware implementations it may prove easier to realize networks with a digital representation of the couplings, of which the simplest is the binary case of $J_{ij} = \pm 1$ or $J_{ij} = 0, 1$. Such networks have been studied previously in models with a specific dependence of the synaptic values on the memories — the 'clipped' Hopfield model[61] and the Willshaw model.[42,62]

In the context of the Gardner approach, the 'Ising' case, $J_{ij} = \pm 1$, was first considered by Gardner and Derrida.[60] The replica symmetric calculation, along the same lines as in the case of the global constraints gives, in the limit $q \rightarrow 1$, a critical storage capacity $\alpha_c = 4/\pi$. This result exceeds the information theoretical bound of $\alpha_c = 1$, indicating that replica symmetry must be broken. Indeed, it has been shown explicitly that the entropy goes to minus infinity on the line $\alpha_c(\kappa)$ obtained in this manner. This adds an additional motivation to study this model as a problem of basic interest for the understanding of the replica method and the mechanisms and scenarios of its breaking. It was investigated recently by Krauth and Mézard,[63] mainly with this goal in mind. First, they pointed out that the limit $q \rightarrow 1$, in which the volume in J-space shrinks to zero, is not appropriate in the present case, because the allowed solutions lie on the corners of the unit hypercube. The critical capacity should be obtained when the volume reaches the minimal value that still contains at least one of these corners. They extended the calculation to one-step replica symmetry breaking and found a critical capacity of $\alpha_c = 0.83$, which is the value at which the replica-symmetric entropy vanishes. The entropy being the (averaged) logarithm of the number of solutions, its zero-crossing is a natural criterion for the actual storage capacity. Let us define $\alpha_{GD}(k)$ as the storage capacity obtained from the criterion $q \to 1$, $\alpha_{AT}(k)$ as the storage capacity obtained on the stability line of replica symmetry, which is the analogue of the Almeida-Thouless line[64] in a spin-glass, and $\alpha_{ZE}(k)$ as the storage capacity obtained from the criterion of zero entropy. It was shown[65] that in a large variety of models with local constraints, replica symmetry breaks down on the line $\alpha_{GD}(k)$ and is stable on the line $\alpha_{ZE}(k)$, namely, $\alpha_{ZE}(k) < \alpha_{AT}(k) < \alpha_{GD}(k)$. Thus α_{ZE} is a consistent estimate of the storage capacity in all such cases.

The calculation of the storage capacity can be extended from the 'Ising' case to larger synaptic depth,[65] when $J_{ij} = \pm 1/L, \pm 2/L, \ldots, \pm 1$. The critical capacity increases rapidly with L and (for $\kappa = 0$) approaches $\alpha_c = 2$ (Fig. 5).

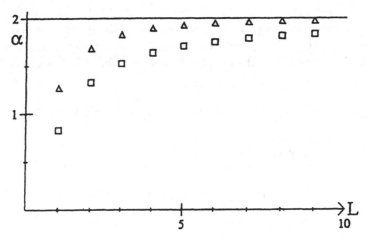

Fig. 5. Critical capacity for discrete synapses as function of the range L. Squares represent results obtained from the zero entropy criterion, and triangles from the $q \to 1$ criterion.

e. Learning with Errors

In the present section we have defined the goal of the learning process as finding synaptic weights for which a set of given patterns satisfies exactly the stability constraints (12). This is a strong requirement and may not be necessary for recognition in an associative memory network. After all, we have seen in the preceding section that in the Hopfield model, the stored patterns are represented by attractors around fixed points which are slightly displaced from the given memories, yielding a small fraction of retrieval errors. Moreover, even if it is possible to satisfy the stability constraints exactly, this is not necessarily the best strategy and there may be advantage in learning with errors.

The fact that a set of patterns are fixed points does not guarantee the performance of the network as a useful realization of an associative memory. The latter depends on the behaviour of the network in the retrieval mode, where the relevant questions are the size of the basins of attraction and the stability of the attractors to synaptic noise. A useful measure, which characterizes this behaviour, is the distribution function of the stability parameters Δ_i^μ

$$\rho(\Delta) = \lim_{\beta \to \infty} \langle \delta(\Delta - \Delta_i^\mu) \rangle_{J,\xi} \tag{42}$$

where the average over J is performed with the Gibbs distribution (Eq. (40)), and the average over ξ requires, again, the replica method.[66,67] The function $\rho(\Delta)$ relates the overlaps, at two consecutive time-steps of synchronous dynamics, of the network states with the stored patterns,

$$m(t+1) = \int d\Delta \rho(\Delta) \mathrm{erf}\left(\frac{m(t)\Delta}{\sqrt{2(1-m^2(t))}} \right) \tag{43}$$

where erf is the error function

$$\mathrm{erf}(x) = \frac{2}{\sqrt{\pi}} \int\limits_0^x dy \exp(-y^2). \tag{44}$$

In networks with assymmetric sparse connectivity[68] in which each neuron is connected on the average to C other neurons, so that $C \approx \log N$, Eq. (43) determines completely the dynamics of overlaps and can be iterated to its fixed point. This allows an analytical study of the retrieval properties in such networks.[69,70] In fully connected networks, the effect of $\rho(\Delta)$ on the basins of attraction can be studied only qualitatively.[66,67]

It is clear that higher weight of the distribution $\rho(\Delta)$ at higher values of Δ implies higher stability. One way to achieve this in a learning algorithm is to increase κ in the stability constraints as much as possible. If one imposes the condition of zero errors in learning, one can use a modification of the perceptron algorithm, known as the minover algorithm,[71] which finds a synaptic matrix for which all the constraints are satisfied with the largest possible κ. It turns out, however, that by increasing κ even further, thereby violating a fraction of the stability constraints, it is possible to improve considerably the retrieval properties.[72]

In the context of the present discussion, it is of interest to consider other 'energy' functions than Eq. (39). A useful modification of the latter is

$$e = (\kappa - \Delta)^n \theta(\kappa - \Delta). \tag{45}$$

This form takes into account the degree by which each of the unsatisfied inequalities (12) is violated. The most interesting cases are $n = 1, 2$. A different 'energy' function appears when these inequalities are replaced by equalities. The natural choice, then, is

$$e = (\kappa - \Delta)^2. \tag{46}$$

The learning process can be described as gradient descent dynamics associated with these 'energy' functions

$$\frac{\partial J_{ij}}{\partial t} = -\frac{\partial E}{\partial J_{ij}}. \tag{47}$$

A discretization of this equation defines a learning algorithm, which leads to the global minimum of E, provided that this function has a sufficiently smooth surface in J-space. Equation (45) with $n = 1$ leads to the perceptron algorithm, Eq. (46) results in the Widrow-Hopf (also called adaline) algorithm,[4] and Eq. (45) with $n = 2$ gives a specific version of the relaxation algorithm.[73] Note that no gradient descent algorithm exists for the 'energy' function of Eq. (39). For a general discussion of the different learning algorithms, we refer the reader to a recent review on the subject by Abbott.[74]

It has been demonstrated[72] that the algorithms associated with the 'energy' functions, defined above, converge to a solution with the minimum energy, as obtained from the replica symmetric calculation leading to Eq. (41), also beyond the critical storage capacity. Although the fixed points, representing the stored patterns, are displaced with a finite shift (fraction of errors) with respect to the patterns, they may have significantly larger basins of attraction than below the critical storage level.

e. Learning with Noise

By taking the limit $\beta \to \infty$ in Eq. (41), we restrict the learning process to zero noise. In the preceding section we emphasized, in a specific context, the advantage of learning with errors. Likewise, in a different context, it may turn out that learning with zero noise may be a poor strategy. This is the case when the goal of learning is to optimize generalization, and will be discussed in Sec. 7. At present, we only wish to point out that the thermodynamic formulation of Sec. 4b allows us to extend the treatment to learning in the presence of noise. The relevant function, then, is the 'free-energy' and a learning algorithm which minimizes the 'free-energy' accepts a change in J according to the probability given by Eq. (15). Learning in the presence of noise leads to a physical realization of the Gibbs distribution in the network parameter space. The resulting Langevin dynamics has been studied for specific learning algorithms.[75,76]

6. Hierarchically Correlated Data and Categorization

The organization of objects with well defined similarity relations in hierarchical structures arises naturally in many cases of data classification and analysis.[77] Such structures would also seem appropriate for some aspects of the organization of semantic information in human memory. It is therefore no wonder that various possible schemes of storing and retrieving hierarchically correlated patterns in associative memory model networks have attracted attention from the early stages of interest in this field.

a. Hierarchical Data Structures

Objects belonging to a hierarchical structure can be viewed as the branches of a tree and the distance between two such objects can be defined as the minimal number of branching points between them. With this definition of distance, any three objects form an isosceles triangle where the two equal sides are longer (or equal) to the third. This property defines what is known as an ultrametric space.[77]

The low-energy states in an SK spin glass are known to span an ultra-metric space, with the distance between two such states defined as $1 - m_{\alpha\beta}$, where $m_{\alpha\beta}$ is the overlap between their equilibrium configurations $\langle S^\alpha \rangle$ and $\langle S^\beta \rangle$.[78] This has motivated a proposal, based on a 'selectionist' learning paradigm,[79] in which the initial state of the network has random synaptic connections and, therefore, has a spin glass energy landscape. Learning proceeds by assigning dynamically low-lying attractors to the learned stimuli and by modifying the synaptic weights to deepen these attractors and to suppress others. This is expected to yield a hierarchically organized memory.

In the standard approach, a memory is built by constructing synaptic weights to embed a given set of patterns. Therefore, in order to study a hierarchically organized memory one has to generate a set of hierarchically correlated patterns. One procedure to construct a simple hierarchical tree of patterns is the following. At the first level of the hierarchy (on a descending scale) one generates p_1 patterns $\{\xi_i^\mu\}$, $\mu = 1, \ldots, p_1$, where every component is chosen independently with the probability

$$P(\xi_i^\mu = \pm 1) = \frac{1}{2}(1 \pm m_1) \tag{48}$$

where we have allowed for correlations between the patterns, already at the first level, induced by the bias m_1. At the second level of the hierarchy, one specifies a new correlation parameter $m_{2\mu}$ and generates from each $\{\xi_i^\mu\}$ pattern , $p_{2\mu}$ descendents $\{\xi_i^{\mu\nu}\}$, $\nu = 1, \ldots, p_{2\mu}$, choosing their components with the probability

$$P(\xi_i^{\mu\nu} = \pm 1) = \frac{1}{2}(1 \pm m_{2\mu} \xi_i^\mu). \tag{49}$$

This rule implies that a component belonging to a pattern descending from the μth ancestor, or prototype, has a higher probability to be equal to ξ_i^μ than to $-\xi_i^\mu$ (since $m_{2\mu} > 0$). The correlations between the patterns in the second generation are given by

$$\begin{aligned} \langle \xi_i^{\mu\nu} \xi_i^{\mu'\nu'} \rangle &= (m_2)^2 \quad (\mu = \mu', \nu \neq \nu') \\ &= (m_1 m_2)^2 \quad (\mu \neq \mu') \end{aligned} \tag{50}$$

where, for simplicity, we assumed that $m_{2\mu}$ is independent of μ. Thus, the patterns are grouped into clusters, or categories, with high correlations between patterns within the same cluster and lower correlations between patterns in different clusters. The correlations of the patterns with the ancestors are

$$\langle \xi_i^{\mu\nu} \xi_i^{\mu'} \rangle = m_2 \qquad (\mu = \mu')$$
$$= (m_1)^2 m_2 \quad (\mu \neq \mu'). \qquad (51)$$

In the case $m_1 = 0$, the patterns are correlated only with the other patterns belonging to their category and only with their own prototype. Note that the case of biased patterns, discussed in Sec. 4c, corresponds to a two-level 'hierarchy' with a single category generated from the ancestor pattern $\{\xi_i\} = \{1, 1, \ldots, 1\}$. This procedure can be continued to produce subsequent generations of the hierarchy.

A different type of hierarchical organization of patterns is achieved when the correlations between the patterns belonging to the same category are localized in a specific group of neurons and not distributed over the entire network.[48,80] An extreme case would be when all the patterns belonging to the same cluster have, say, the first N_1 bits in common, the remaining strings of $N - N_1$ bits, characterizing the different patterns, are uncorrelated. In the case of patterns belonging to different clusters, the strings of the first N_1 bits are also uncorrelated. Such a hierarchical organization of data leads to a novel network architecture by separating the network into two functionally different parts.[80] The distinction between the two parts is even more apparent if, in addition, one introduces a different structure (like dilution, symmetry, overall strength) of synaptic connections within the two groups of neurons and between them. This scheme can easily be extended to hierarchies of more levels. A higher level category, which contains several clusters, can be identified by, say, the first N_2 bits of string N_1, and so forth.

b. Storage of Hierarchical Data Structures

In the present section we review two schemes of constructing an associative memory with hierarchically correlated patterns. Both belong to the class of models in which the synaptic efficacies are prescribed explicitly, and both can be treated analytically by the same method which was used to analyze the Hopfield model.

Let us consider a two-level hierarchy in which the prototypes (ancestors) are uncorrelated and unbiased, and have a common overlap m with their offsprings (descendents). The patterns $\{\xi^{\mu\nu}\}$ and the prototypes $\{\xi^{\mu}\}$ can be stored in a network with synaptic efficacies[38]

$$J_{ij} = \frac{1}{N} \sum \xi_j^{\mu} \xi_j^{\mu} + \frac{1}{N\Delta} \sum_{\mu,\nu} (\xi_i^{\mu\nu} - m\xi_i^{\mu})(\xi_j^{\mu\nu} - m\xi_j^{\mu}). \qquad (52)$$

This is a slight modification of an earlier proposal along these lines.[47] The phase diagram of this model has been analyzed in detail.[38] When $\Delta = \Delta_0 = 1 - m^2$, the descendent patterns and their ancestors are equivalent. They are embedded with the same energy and are destroyed at the same level of noise (same transition temperature T_M). The storage capacity is again $\alpha \approx p/N = 0.14$, where p is the total number of stored patterns — the prototypes and their descendents. The capacity (at $T = 0$) is unchanged when Δ varies in the range $\Delta_0 \leq \Delta < \Delta_0(1 + m)/m$, but the equivalence between the prototypes and the descendent patterns is lifted and the attractor states corresponding to the prototype configurations have a lower energy and are destroyed at a temperature T_{M1}, exceeding the temperature T_{M2} at which the states representing the descendent patterns become unstable.

Increasing the synaptic noise to a level corresponding to a temperature in the range $T_{M1} < T < T_{M2}$ leads to a phase in which the memory of the specific objects is lost, but the classes to which they belong are still recognized. The same situation can be achieved by introducing damage into the network through the breaking of synaptic bonds, which has nearly the same effect as fast synaptic noise.[61] This reduction to prototype may be identified as an additional aspect of the theme of categorization, alongside with prototype formation and prototype dominance, mentioned earlier (Sec. 3c). At this point it is natural to make contact with a well-known syndrome called prosopagnosia, which is associated with a lesion in the central visual system. Prosopagnosia generally refers to an impairment in face recognition, but is often accompanied by a more general difficulty to identify one individual among others, which are visually similar and therefore belong to the same class.[81] On the other hand, the category which describes the class is correctly recognized. The relation between the effect of random synaptic lesions in hierarchically organized memory networks and prosopagnosia has been elaborated by Virasoro.[24]

A different scheme for constructing an associative memory, has been proposed,[49] in which the patterns belonging to different levels of the hierarchical tree are stored in successive networks of same size (number of neurons). In the case of the same data structure as above, the prototypes are stored in the first network by the ordinary Hebb rule (first term of Eq. (52)) and the individual patterns are embedded in a second network with synaptic efficacies given by the second term in Eq. (52). There are one-way connections from every neuron in the first network to one of the neurons in the second network, transferring the activity in the former network to the latter. The process of retrieval proceeds as follows. One of the descendent patterns $\{\xi^{\mu\nu}\}$ is presented as an external stimulus to both networks. Since it has a significant overlap with its prototype pattern $\{\xi^{\mu}\}$, it is assumed that it is within the basin of attraction of the latter. The pattern $\{\xi^{\mu}\}$ is, thus, retrieved in the first network, and the neurons in the second network receive a signal, acting as an external field, of the form $h_i = h\xi_i^{\mu}$. It has been shown that for a wide range of values of the synaptic strength h, the effect of the signal from the first network is to constrain the dynamics in the second network to a subspace spanned by the cluster of patterns belonging to the category $\{\xi^{\mu}\}$. This constraint, together with the specific form of the J_{ij}'s, guarantees that the specific pattern is retrieved in the second network. In such an architecture, recognition of global characters precedes and helps recognition of details.

A malfunctioning of the recognition process, resembling the prosopag-nosia syndrome, arises naturally also in this scheme. Suppose that one introduces a lesion which breaks the connection between the two networks. Then, the class will still be recognized by the first network. The second network is overloaded by the specific objects and the retrieval of any one of them is possible only with the help of information on the corresponding class, arriving from the first network.

The last model of a hierarchical memory can be used to construct a hierarchical tree of networks. Let us consider the case of several different hierarchical trees of patterns. The individual patterns of each of them are stored in different networks, but they can share the same network for their ancestor states. This network has efferent connections to all the networks storing the specific patterns. Without carrying the analogy too far, such a structure is reminiscent of the hierarchical organization of cortical areas in the visual system of the macaque monkey, where the neurons in the lowest areas $(V1, V2)$ respond to several domains of information, such as shape and motion, while the response in the higher levels is more specific.[82]

c. Categorization

The models of the storage and retrieval of hierarchical data structures, described in the preceding section have several appealing features, but they are not quite satisfactory. In a sense, they are designed and carefully tuned to produce the expected results. In particular, the patterns corresponding to the categories have to be known *a priori* and have to be built-in explicitly into the synaptic efficacies. In a more realistic model of the process of categorization, we would rather expect the spontaneous appearance of attractors, describing the categories, as a byproduct of learning a series of similar objects (the prototype formation, mentioned above).

This is, indeed, the property of a typical synaptic matrix, which guarantees that a set of given patterns $\{\xi^{\mu\nu}\}$, with correlations defined by Eq. (50), are fixed points of the network dynamics. In the context of Sec. 5, this means that they satisfy Eqs. (12) and (13). In the course of the calculation of the storage capacity, or the density of local stabilities (Eq. (42)), there appear new order parameters

$$M^{\mu} = \xi_i^{\mu} \sum_j J_{ij} \xi_j^{\mu} \tag{53}$$

where, like in Sec. 5, the sites i are equivalent and independent, and, therefore, we have suppressed, on the l.h.s., the dependence on i. The configurations $\{\xi_i^{\mu}\}$ are given by

$$\xi_i^{\mu} = \text{sign}\left(\sum_{\nu} \xi_i^{\mu\nu}\right). \tag{54}$$

It is easy to check that their overlaps with the stored pattern are given by Eq. (51) and, hence, they are precisely the ancestor patterns which characterize the different classes. In the case of a single cluster, the ancestor pattern can be gauge-transformed to $\{1, 1, \ldots, 1\}$ and M becomes the order parameter, which appears in the treatment of biased patterns (Eq. (37)). It was mentioned there that, typically, $M > 0$. This result is also true in the case of several clusters and typically $M^{\mu} > 0$, indicating that all the ancestor patterns are also fixed points of the network dynamics.

The main point is that the stability of these patterns was not imposed explicitly in the calculation of the storage capacity or in the learning algorithm, which was used to derive a particular set of synaptic weights which guarantee the stability of the patterns $\{\xi^{\mu\nu}\}$. Moreover, the prototype patterns $\{\xi^{\mu}\}$ are, typically, more stable than the patterns themselves. This can be demonstrated analytically by treating separately the sites at which $\xi_i^{\mu\nu}\xi_i^{\mu} > 0$ and those at which $\xi_i^{\mu\nu}\xi_i^{\mu} < 0$, and calculating the distribution of local stabilities (Eq. (42)) for the two types

$$\rho_{\pm}(\Delta) = \lim_{\beta\to\infty} \langle\delta(\Delta_i^{\mu\nu} - \Delta)\theta(\pm\xi_i^{\mu}\xi_i^{\mu\nu})\rangle_{J,\xi}. \tag{55}$$

The average stability of the two types of sites is given by

$$\langle\Delta_{\pm}\rangle = \int \Delta\rho_{\pm}(\Delta)d\Delta. \tag{56}$$

It has been shown[24] that $\langle\Delta_{+}\rangle$ is larger than $\langle\Delta_{-}\rangle$, indicating that, indeed, the prototype patterns $\{\xi^{\mu}\}$ are more stable than the descendent patterns $\{\xi^{\mu\nu}\}$. This can, again, be demonstrated by introducing random lesions with the same effect as described in the last section.

7. Generalization

In Sec. 5 we defined the task of learning in neural network models as finding synaptic weights for which a set of given configurations become fixed points of the network dynamics (in attractor networks) or which associate desired outputs with a set of given inputs (in feedforward networks). In the present section the emphasis is on learning specific input-output associations, not for their own sake, but as examples for the inductive inference of a hidden rule. Inductive inference (or learning from examples) is at the heart of a large scope of artificial intelligence programs, expert systems and human cognitive activities. It has been formalized in the mathematical framework of computational learning theory.[83] The learning of tasks which can be formulated as a function, relating N-component inputs to N'-component outputs, can most naturally be implemented in feedforward networks.

For a neural network, with a given set of constrained resources (size, structure, synaptic variability), several questions may come to mind: Is a given task 'learnable', e.g., does there exist an allowed configuration that produces the correct output for any input? Assuming that a given task is learnable: What is the best learning strategy to optimize the generalization error for a given size of the learning set? And even, what can best be achieved for an 'unlearnable' task?

The problem of learning and generalization by layered networks has recently been discussed extensively with emphasis on phase space, entropy and thermodynamic considerations.[84-86] The simplest layered network is the perceptron. In this case the Statistical Mechanics analysis of learning, based on the work of Gardner, can be applied to study also the generalization ability. This activity is only at its beginning, but interesting insights and results have already been obtained, and it seems that it will be possible to apply these methods also to multi-layer networks. To give the flavour of this development, we shall briefly review some of the basic results.

a. Learning a Classification Task

Let us start with a simple example of classification. To prepare the grounds, we shall first consider the problem of learning a given set of input-output associations on a perceptron with N input and N' output units. The thermodynamic limit implies that $N \to \infty$, but N' can remain finite, and we shall take, for convenience, $N' = 1$. The bits in the input patterns are chosen at random with a bias m_{in} (Eq. (21)) and the output bit is chosen with the same probability, but with a bias parameter m_{out}. When $m_{in} = 0$, then the critical storage capacity is $\alpha_c = 2$ for all values of m_{out}. When $m_{in} \neq 0$, then α_C is a monotonically increasing function of m_{out}. An infinite capacity can be achieved in the limit $m_{out} \to 1$. In this limit there is a unique output state for all inputs with bias $m_{in} > 1$. It has been suggested[87] that this result can be used to distinguish between input patterns according to the sign of their bias m, because by symmetry all patterns with a negative bias will give $m_{out} = -1$, and that the network can be trained to perform this task through the presentation of a finite number of examples. The fact that the capacity is infinite means that this task is learnable. Learnability in a perceptron with a single output unit is equivalent to the geometrical property of linear separability.[3,58]

This simply result can be extended[88] to the classification of input patterns according to their proximity to a finite number M of given uncorrelated prototype patterns $\{\xi_i^\mu\}(\mu = 1, \ldots, M)$. Each of these prototypes is associated with an output bit σ^μ. The synaptic weights are adjusted, by some learning procedure, so as to satisfy the conditions

$$\sigma^\mu = \text{sign}\left(\sum_i J_i \xi_i^{\mu\nu}\right), \quad (\nu = 1, \ldots, p).\tag{57}$$

The input patterns $\{\xi_i^{\mu\nu}\}$ are generated from the prototypes using the probability distribution of Eq. (49) with a parameter m_{in} which measures their overlap with the corresponding prototype. After a training session involving a sufficiently large number of examples, every new pattern characterized by the same or larger overlap with one of the patterns will be mapped by the perceptron onto the corresponding output.

This problem has recently been extended to the case when the number of prototypes scales with N ($M = \alpha N$ and $N \to \infty$) and the generalization process has been analyzed within a statistical mechanics formulation.[89] The training process can be characterized by a cost-function of the form of Eq. (38), which is now more conveniently referred to as the training error. The training error is naturally measured by the number of misclassified examples:

$$E_r(J) = \sum_{\mu=1}^{M} \sum_{\nu=1}^{P} \theta \left(-\sigma^\mu \sum_i J_i \xi_i^{\mu\nu} \right) . \tag{58}$$

The training can, in general, be described as a stochastic process with a noise parameter β (Sec. 5e), leading to a Gibbs distribution in configuration space (Eq. (40)) with $E_T(J)$. The average training error per example is

$$e_t = \frac{1}{pM} \langle E_T(J) \rangle_{J,\xi} \tag{59}$$

where the average over J is performed with the Gibbs measure, and the average over ξ indicates averaging over the distribution of the prototypes and the learned examples. The interesting question is how many examples have to be learned so that a new input pattern, in the neighborhood of one of the prototypes, will be classified correctly with a probability $1 - e_g$? This defines the generalization error e_g, which in this case is related to the distribution of the local stabilities (Eq. (42)) generated by the prototypes.

Let us summarize the main results of Ref. 89, emphasizing the conceptually important features. The behavior of the generalization error depends on the training noise β and on the number of examples p. Training without errors is possible, at $\beta \to \infty$, up to a critical p_c, which depends on α and m_{in}. The generalization error decreases with increasing p, but the interesting feature is that the minimum is obtained before the critical value, p_c, is reached. This indicates that the addition of examples does not necessarily improve the performance on new inputs. To improve generalization one has to allow errors in learning. This can be achieved in two ways. One is to increase the number of examples beyond p_c, but still learn without noise, namely with $\beta \to \infty$. A better strategy, which results in a faster decrease of the generalization error towards a minimal critical value e_c, is to learn with noise (Sec. 5e). Rather than learning at a fixed level of noise it is possible to adjust β so as to keep the training error e_t fixed. It turns out that the best performance (at large p) is achieved when $e_t = e_c$.

b. The Reference Perceptron Problem

Let us consider a perceptron with N input units and a single output unit, and with given weights R_i. The letter R indicates that it plays the role of a reference (or 'teacher') perceptron. Such a perceptron determines a Boolean function defined on the space of N binary inputs S_i:

$$\sigma \equiv \sigma(R, S) = \text{sign} \left(\sum_i R_i S_i \right) \tag{60}$$

The question is how effectively this function can be implemented on a perceptron of similar architecture by adjusting its weights on the basis of a set of $p = \alpha N$ randomly chosen examples $\{S^\mu\}$. This problem was first introduced and analyzed numerically for binary weights $R_i = \pm 1$ in Ref. 90. The weights of the trained perceptron can be given by a specific expression, like the Hebb rule or the projection rule, or can be determined dynamically by a learning algorithm. We refer to both cases as the training process, the result of which is a perceptron with weights J_i. The training process is, again, characterized by a training error which is naturally measured by the number of examples in disagreement with the reference perceptron

$$E_T(J) = \sum_{\mu=1}^p \theta(-\sigma(R, S^\mu)\sigma(J, S^\mu)) \,. \tag{61}$$

The generalization error is defined as this measure averaged over the entire input space

$$e_g(J) = 2^{-N} \sum_{\{S\}} \theta(-\sigma(R, S)\sigma(J, S)) \,. \tag{62}$$

A straightforward calculation gives

$$e_g(J) = \frac{1}{\pi} \arccos(\rho) \,, \tag{63}$$

where ρ is the cosine of the angle between the vectors $\mathbf{R}$ and $\mathbf{J}$. The parameter ρ can be easily calculated for the Hebb rule and the projection rule.[91] The interesting result is that, although the projection rule yields a significantly smaller training error, a perceptron constructed by the Hebb rule generalizes better. The asymptotic behavior is in both cases $e_g \sim \alpha^{-1/2}$. This does not mean that a larger training error, in general, implies better generalization. When the weights are derived by a learning algorithm which gives the best possible stabilities for the learned examples (like the minover algorithm[71]), the training error is zero. Nevertheless, it was found[92] that in such a perceptron the generalization error asymptotically decreases faster than in the two previous cases, namely $e_g \sim \alpha^{-1}$.

A different approach is to analyze the generalization properties independently of a specific learning algorithm.[93,94] In this case one gets typical results averaged over all the possible configurations of weights which give the same training error. Likewise, it is of interest to include the possibility of training with noise. The average training error per pattern is then defined, as in the previous section, by averaging Eq. (61) over the Gibbs distribution in $\mathbf{J}$-space and over the realization of the learned patterns. The generalization error is given by the same kind of average of Eq. (62).

It has been shown[94] that in the case of continuous weights, the generalization error decreases asymptotically as $1/\alpha$. A qualitatively different behavior is found for the binary perceptron. In this case[94,95] a sharp transition is found at low β (high temperature) for a specific value of $\alpha\beta$. Above this value, a training procedure starting from random initial weights (giving $\rho \sim 0$) converges rapidly to $\rho = 1$. This transition to perfect generalization indicates that there are no networks with $e_g > 0$ which are consistent with the learned examples.

An interesting modification of the reference perceptron is to allow external noise in the examples. This is achieved when the answers during the training session are provided by a perceptron with weights which are Gaussian distributed around those of the reference perceptron. It is found that above a certain number of examples, depending on the external noise, the best strategy is to add internal noise during the learning process. This demonstrates again, as in the previous section, that optimal generalization involves an interplay between the size of the training set and the training noise.

c. The Contiguity Problem

Let us finally discuss the recognition of a simple geometrical feature of the input patterns. The task is to count the number of contiguous domains of, say, the +1 bits, or to distinguish patterns with a number of domains, exceeding some given value, from all the others. The counting of domains is equivalent to the counting of edges between them and can be implemented by a network of overlapping edge detectors, each of which is connected to two adjacent input sites. Edge detection is an elementary component of early visual processing, and it is no wonder that the contiguity problem has played an important role in the history of adaptive neural networks[3] and automatic learning. The dynamical process under which a network of edge detectors learns to solve contiguity problems has recently been analyzed within the framework of the statistical mechanics of learning and generalization.[96] This study is the first example of the application of these methods to generalization by a (admittedly simple) two-layer network model.

There exists a simple 'engineer's solution' to implement the task of counting the number of domains on a two-layer network with a single output unit. The patterns $S_i = \pm 1$ are presented at the input layer of $N + 1$ elements with periodic boundary conditions, $S_1 = S_{N+1}$. Each pair of adjacent input bits is connected to a hidden unit. The value of the latter is determined by

$$\sigma_i^h = \text{sign}(J_{iR}S_{iR} + J_{iL}S_{iL} - 1) \tag{64}$$

where the subscripts R, L indicate the input unit on the right and left of the hidden unit i. If $J_{iR} = 1$ and $J_{iL} = -1$, such a hidden unit detects an 'edge' between +1 bits on the right and −1 bits on the left. The number of domains is simply the number of such edges. To count the number of edges, one connects all the hidden units to a linear output unit

$$\sigma = \frac{1}{2}\left(\sum_{i=1}^{N} \sigma_i^h + N\right). \tag{65}$$

The analysis[96] of the learning of the task of counting domains with this architecture by adjusting only the values of the binary weights reveals a novel scenario. It was shown that the generalization error decreases exponentially as $e_g \sim \exp(-2\beta\alpha)$, where α is the number of examples per weight. A modification of the task from counting the number of domains to recognizing networks with a number of domains larger than a certain value N_0 results in a different behavior. This new task is accomplished by changing the output unit to a threshold device

$$\sigma = \text{sign}\left(\frac{1}{2} \sum_{i=1}^{N} \sigma_i^h + \frac{1}{2}N - N_0 \right). \tag{66}$$

In this case, one finds for all values of the training noise a discontinuous transition, at a critical α, from a high value of e_g to perfect generalization. These results are reproduced convincingly in numerical simulations.

8. Discussion – Issues of Relevance

a. The Notion of Attractors and Modes of Computation

In order to bring the neural network models into some biological perspective, it is useful to evoke the distinctions, introduced by Fodor,[97] among the natural neural structures. Fodor distinguishes between three types: Transducer systems (sensory captors like the retina or the cochlea), input systems (involved in the preprocessing of sensory data) and central cognitive systems (higher brain areas).

Much of the work in neurophysiology and psychophysics has focused on the study of transducer and input systems, for the good reason that they are more reproducible and easier to interpret.[98] The emphasis has been on proving that many perceptual phenomena are indeed bottom-up, and encapsulated, and therefore not influenced by the elusive higher brain areas. Quite naturally, the neural networks selected to model input systems are of the feedforward type.

In the present stage of brain modeling, the main biological relevance of attractor neural networks is for the understanding of higher brain functions. The idea is that in higher areas (also called associative areas), which are the sites for cognitive functions such as semantic recognition and association, judgement and decisions, something different from the feedforward mode of computation is called for. In this context, the statistical physics of neural networks has provided a biologically relevant suggestion, with the notion of computation-with-attractors and the introduction of a family of models that allows for a qualitative and progressively quantitative confrontation with neurobiological data.

The notion of computation with attractors can be related to the ideas of Gestalt psychologists, who began, early in this century, to stress many puzzling features of perception and thought, such as the segmentation of stimuli into objects and categorization of objects into classes. They showed in essence that percepts may contain much more, and also less, than sensory data. They conjectured that the basic mechanism of perception is the convergence towards minimum energy configurations of electric field distributions in the brain. The primitive picture may be seen nowadays as an early naive percursor of the scientific process that has led to the establishment of the modern notion of computation with attractors.[99]

b. The Nature of Attractors

In a simple interpretation of the attractor state a neuron is supposed to fire either strongly (close to the maximum possible rate) or weakly (close to the background firing level). A systematic study of the occurrence of high firing rates in association areas of the cortex[100] indicates that, althought occasional bursts are recorded, they are too rare and uncorrelated to provide evidence for the existence of attractor states, at least not in their simple interpretation. On the other hand, simultaneous recording from groups of neurons shows correlated activity in several neurons, though at rates of 10–15 sec^{-1}, much lower than the maximum possible. Moreover, the firing rates vary under different conditions and in different areas of the cortex.

These experimental findings have led to a profound modification of the models, focusing the attention on low firing rates with tunable activity.[101,102] However, the new models preserve the main conceptual framework, i.e., the computation in terms of a flow to associative attractors in the presence of dynamic noise. The two models, quoted here, have several common features which bring them much closer to neurophysiological reality:

a) The neurons in the network are explicitly separated into excitatory (typically the pyramidal cells) and inhibitory (typically the stellate cells) neurons, reflecting the different morphology and physiology of the two types.

b) Memory is stored in the synaptic connections between the excitatory neurons, reflecting the long range connectivity of the pyramidal cells. The role of inhibition is to control the overall activity of the network.

c) The neurons are typically below threshold and emit spikes stochastically due to the noise in the network, the spiking probability being modulated by the overall inhibition. The role of the synaptic connectivity between the excitatory neurons is to bring the neurons, which are active in a given pattern, closer to the firing threshold than the other neurons.

d) Since neurons spike stochastically, the individual spike times of the different neurons participating in a memory pattern are not correlated, but the enhanced rates are.

c. Temporal versus Spatial Coding

In the models of self-coupled networks, described in this review, the average activity of the individual neurons is the only variable used to encode information. An attractor state consists of two groups of neurons, which are fixed during its lifetime and are distinguished by their activity amplitude. Thus, coding by attractors is purely spatial.

It is commonly admitted that in the peripheral areas the intensity of an external stimulus is coded by the average activity (firing frequency) of sensor neurons. Likewise, several suggestions have been made about the information which might be coded by the neuronal activity amplitude in the cortex. But, meanwhile, it was argued[103,104] that spatial coding cannot be sufficient for the processing of sensory information and that temporal correlations between the activities of individual neurons are required for the linking of sensory inputs between disconnected receptive fields or between different sensory features and modalities (e.g., visual and auditory). Such linkage seems to be essential to segment different objects in a complex scene and to separate them from the background. A specific neural model for the 'cocktail-party effect' was proposed along these lines.[103] It is common experience that one can focus on, and follow, a particular conversation despite the fact that the signal is weaker than that of the background cross-talk at a cocktail party. It was suggested that the relevant stimulus is identified by a delayed coincidence in the firing patterns of neurons receiving input signals from the two ears.

These ideas have recently received experimental support from the discovery of oscillatory neural responses in the primary visual cortex of anesthetized cats.[105,106] This discovery raised anew the question of the neural code and has had already such an impact on the neurobiological community that it is worth mentioning the basic findings. The activity of neurons responding to a moving bar with a specific orientation has an oscillatory component with an average period of ≈ 20–30 ms, which appears to be the same for different neurons and for different orientations of the bar. The oscillating component of neurons receiving inputs from the same receptive field are synchronized when responding to a single oriented bar. Neurons receiving inputs from distant receptive fields oscillate coherently only when responding to two bars of the same orientation and moving in the same direction (or, preferably, to a single bar extending over the two receptive fields).

Temporal correlations introduce the phase as an additional variable describing neural activity and open new horizons for neural network models. Models of networks in which the neurons are phenomenologically represented by oscillators have been proposed.[107,108] These models represent yet another contribution of physics to neural network science by making contact with previous work on the collective behavior of randomly coupled phase oscillators, in which the relevant questions of frequency and phase locking have been studied.[109] A different direction, motivated by the discovery of oscillations and which avoids phenomenological oscillating elements, consists in modifications of the existing attractor network models, even beyond the model described in Refs. 101, 102, which bring them still closer to biological reality.[110,111] Such models are difficult to treat analytically and have to be studied by numerical simulations. It is shown that the activity of neurons participating in the retrieval of a stored pattern may exhibit clear oscillations, similar to the ones observed in experiment. This behavior is based on the fact that an integrate-and-fire neuron, in the presence of persistent input, is an oscillator.

Let us close this section by mentioning that the oscillations in neural activity, of a period of $\approx$ 20–30 ms, have recently been conjectured to play an essential role in neural mechanisms underlying short-term memory and consciousness.[112]

Acknowledgements

The idea to write a review article emphasizing the contributions of physics to the field of neural network models has its origins in 1987, when we participated in a workshop on this subject at the Institute for Advanced Studies at the Hebrew University of Jerusalem. We are grateful to the Institute for providing an ideal framework and atmosphere, and to our colleagues at the workshop — physicists, neurobiologists, computer scientists and psychologists — for helping us to crystallize our ideas. One of us (H. G.) is grateful to the Laboratoire de Physique of the Ecole Normale Supèrieure and, in particular, to his colleagues of the Laboratoire de Physique Statistique, for the warm hospitality extended to him during a visit when this article was completed.

References

1. W. S. McCulloch and W. A. Pitts, *Bull. Math. Biophys.* **5**, 115 (1943).
2. D. O. Hebb, *The Organization of Behavior* (Wiley, 1949).
3. F. Rosenblatt, *Principles of Neurodynamics* (Spartan, 1961); M. Minsky and S. Papert, *Perceptrons* (MIT Press, 1988).
4. B. Widrow and M. E. Hoff, IRE WESCON, *Convention Report* **4**, 96 (1960).
5. S. Amari and K. Maginu, *Neural Networks* **1**, 63 (1988) and references therein.
6. E. R. Caianiello, *J. Theor. Biol.* **2**, 204 (1961).
7. S. Grossberg, *Neural Networks* **1**, 17 (1988) and references therein.
8. T. Kohonen, *Self-Organization and Associative Memory* (Springer, 1984 and 1989).
9. V. I. Kryukov, in *Stochastic Cellular Systems: Ergodicity, Memory, Morphogenesis* (Manchester Univ. Press, 1990).
10. G. Palm, *Neural Assemblies, an Alternative Approach to Artificial Intelligence* (Springer, 1982).
11. B. G. Cragg and H. N. V. Temperley, *Electroenceph. Clin. Neurophysiol.* **6**, 85 (1954).
12. W. A. Little, *Math. Biosci.* **19**, 101 (1974); W. A. Little and G. L. Shaw, *ibid.* **39**, 281 (1978).
13. J. J. Hopfield, *Proc. Natl. Acad. Sci. (USA)* **79**, 2554 (1982); **81**, 3088 (1984).
14. D. J. Amit, H. Gutfreund, and H. Sompolinsky, *Ann. Phys. N.Y.* **173**, 30 (1987).
15. E. Gardner, *J. Phys.* **A21**, 257 (1988).
16. D. J. Amit, *Modeling Brain Function* (Cambridge Univ. Press, 1989).
17. V. Braitenberg, *On the Texture of Brains* (Springer, 1977).
18. P. S. Churchland, *Neurophilosophy. Toward a Unified Science of the Mind-Brain* (MIT Press, 1986).
19. P. Peretto, *The Modeling of Neural Networks* (Cambridge Univ. Press, 1991).
20. For a review see: H. C. Tuckwell, *Introduction to Theoretical Neurobiology* (Cambridge Univ. Press, 1988).
21. L. F. Abbott and T. B. Kepler, in *Proc. XI Sitges Conf. on Neural Networks*, Sitges, Spain, June 1990.
22. G. L. Shaw and R. Vasudevan, *Math. Biosci.* **21**, 207 (1974).
23. D. E. Rumelhart and J. L. McClelland, *Parallel Distributed Processing*, Vols. 1 and 2 (Bradford Books, Cambridge, MA, 1986).
24. M. A. Virasoro, *Europhys. Lett.* **7**, 293 (1988).
25. S. Kirkpatrick and G. Toulouse, *J. Physique* **46**, 1277 (1985).

26. M. Mézard, G. Parisi, and M. A. Virasoro, *Spin Glass Theory and Beyond* (World Scientific, 1987), Chaps. VII–IX.
27. S. Kirkpatrick, C. D. Gelatt Jr., and M. P. Vecchi, *Science* **220**, 671 (1983).
28. J. J. Hopfield and D. W. Tank, *Biol. Cybern.* **52**, 141 (1985).
29. G. Weisbuch and F. Fogelman-Soulié, *J. Physique Lett.* **2**, 337 (1985).
30. S. Kirkpatrick and D. Sherrington, *Phys. Rev.* **B17**, 983 (1978).
31. Ref. 25, Chap. I.
32. S. F. Edwards and P.W. Anderson, *J. Phys.* **F5**, 965 (1975).
33. A. Crisanti, D. J. Amit, and H. Gutfreund, *Europhys. Lett.* **2**, 337 (1986).
34. D. C. Mattis, *Phys. Lett.* **56A**, 421 (1976).
35. R. Meir and E. Domany, *Phys. Rev.* **A37**, 2660 (1988).
36. Y. Stein, Ph.D. Thesis, The Hebrew University of Jerusalem.
37. G. Parisi, *J. Phys.* **A19**, L675 (1986).
38. M. V. Feigelman and L. B. Ioffe, *Int. J. Mod. Phys.* **B1**, 51 (1987).
39. J. A. Hertz, G. Grinstein, and S. Solla, in *Proc. Heidelberg Colloquium on Glassy Dynamics*, ed. J. L. van Hemmen and I. Morgenstern (Springer, 1987), p. 538.
40. A. Cristanti and H. Sompolinsky, *Phys. Rev.* **A36**, 4922 (1987); *ibid.* **A37**, 4865 (1988).
41. P. Spitzner and W. Kinzel, *Z. Phys.* **B77**, 511 (1989); T. Pfenning, H. Rieger, and M. Schreckenberg, *J. Physique*, in press.
42. D. J. Willshaw, O. P. Buneman, and H. C. Longuet-Higgins, *Nature* **222**, 960 (1969).
43. J. Buhmann, R. Divko, and K. Schulten, *Phys. Rev.* **A39**, 2689 (1989).
44. M. V. Tsodyks and M. V. Feigelman, *Europhys. Lett.* **6**, 101 (1988).
45. C. J. Perez Vincente and D. J. Amit, *J. Phys.* **A22**, 559 (1989).
46. D. J. Amit, H. Gutfreund, and H. Sompolinsky, *Phys. Rev.* **A35**, 2293 (1987).
47. N. Parga and M. A. Virasoro, *J. Physique* **47**, 1857 (1986).
48. V. Dotsenko, *Physica* **140A**, 410 (1986).
49. H. Gutfreund, *Phys. Rev.* **A37**, 570 (1988).
50. N. Sourlas, *Europhys. Lett.* **7**, 749 (1988).
51. L. Personnaz, I. Guyon, and G. Dreyfus, *J. Physique Lett.* **46**, L359 (1985).
52. I. Kanter and H. Sompolinsky, *Phys. Rev.* **A35**, 380 (1987).
53. S. Diederich and M. Opper, *Phys. Rev. Lett.* **58**, 949 (1987).
54. J.-P. Nadal, G. Toulouse, J. P. Changeux, and S. Dehaene, *Europhys. Lett.* **1**, 535 (1986).
55. M. Mézard, J.-P. Nadal, and G. Toulouse, *J. Physique* **47**, 1457 (1986).
56. G. Parisi, *J. Phys.* **A19**, L617 (1986).
57. J.-P. Nadal, G. Toulouse, J. P. Changeux, and S. Dehaene, in *Computer Simulations in Brain Science*, ed. R. M. J. Cotterill (Cambridge Univ. Press, 1987), p. 221.
58. T. M. Cover, *IEEE Trans. Electron. Comput.* **14**, 326 (1965).
59. S. Venkatesh, Ph.D. Thesis, California Institute of Technology, 1986.
60. E. Gardner and B. Derrida, *J. Phys.* **A21**, 271 (1988).

61. H. Sompolinsky, *Phys. Rev.* **A34**, 2571 (1986); in *Proc. Heidelberg Colloquium on Glassy Dynamics*, ed. J. L. van Hemmen and I. Morgenstern (Springer, 1987), p. 485.

62. D. Golomb, N. Rubin, and H. Sompolinsky, *Phys. Rev.* **A41**, 1843 (1990).

63. W. Krauth and M. Mézard, *J. Physique* **50**, 3057 (1989).

64. J. R. de Almeida and D. J. Thouless, *J. Phys.* **A11**, 983 (1978).

65. H. Gutfreund and Y. Stein, *J. Phys.* **A23**, 2613 (1990).

66. T. B. Kepler and L. F. Abbott, *J. Physique* **49**, 1657 (1988).

67. W. Krauth, J.-P. Nadal, and M. Mézard, *J. Phys.* **A21**, 2995 (1988).

68. B. Derrida, E. Gardner, and A. Zippelius, *Europhys. Lett.* **4**, 167 (1987).

69. E. Gardner, *J. Phys.* **A22**, 1966 (1989).

70. D. J. Amit, M. R. Evans, H. Horner, and K. Y. M. Wong, *J. Phys.* **A23**, 3361 (1990).

71. W. Krauth and M. Mézard, *J. Phys.* **A20**, L745 (1987).

72. M. Griniasty and H. Gutfreund, *J. Phys.* **A** (in press).

73. L. F. Abbott and T. B. Kepler, *J. Phys.* **A22**, L711 (1989).

74. L. F. Abbott, *Network* **1**, 105 (1990).

75. J. A. Hertz, A. Krogh, and G. A. Thorbergson, *J. Phys.* **A22**, 2133 (1989).

76. W. Kinzel and M. Opper, in *Physics of Neural Networks*, ed. J. L. van Hemmen, E. Domany and K. Schulten (Springer, 1989).

77. R. Rammal, G. Toulouse, and M. A. Virasoro, *Rev. Mod. Phys.* **58**, 765 (1986).

78. Ref. 25, Chap. IV.

79. G. Toulouse, S. Dehaene, and J. P. Changeux, *Proc. Natl. Acad. Sci. (USA)* **83**, 1695 (1986).

80. C. Fassnacht and A. Zippelius, preprint (1990).

81. A. R. Damasio, H. Damasio, and G. W. van Hessen, *Neurology (N.Y.)* **32**, 331 (1982); E. Warrington and T. Shallice, *Brain* **107**, 829 (1984).

82. D. C. van Essen and J. H. R. Maunsell, *Trends in Neurosciences* **6**, 370 (1983).

83. L. A. Valiant, *Comm. ACM* **27**, 1134 (1984); *Phil. Trans. Roy. Soc. London* **A312**, 441 (1984).

84. S. Patarnello and P. Carnevali, *Europhys. Lett.* **4**, 503 (1987); *ibid.* **4**, 1199 (1987).

85. N. Tishby, E. Levin, and S. Solla, in *Proc. Int. Joint Conf. on Neural Networks* Vol. 2, p. 403; E. Levin, N. Tishby, and S. Solla, in *Proc. Second Annual Workshop on Computational Learning Theory, COLT-89*, ed. R. Rivest, D. Haussler, and M. K. Warmuth (Morgan Kaufmann, CA, 1989), p. 245.

86. J. Denker, D. Schwartz, B. Winter, S. Solla, R. Howard, L. Jackel, and J. Hopfield, *Complex Systems* **1**, 877 (1987).

87. W. Krauth, M. Mézard, and J. P. Nadal, *Complex Systems* **2**, 387 (1988).

88. P. del Giudice, S. Franz, and M. A. Virasoro, *J. Physique* **50**, 121 (1989).

89. D. Hansel and H. Sompolinsky, *Europhys. Lett.* **11**, 687 (1990).

90. E. Gardner and B. Derrida, *J. Phys.* **A22**, 1983 (1989).

91. F. Vallet, *Europhys. Lett.* **8**, 747 (1989); F. Vallet, J. G. Cailton, and P. Refrégier, *ibid.* **9**, 315 (1989).

92. M. Opper, W. Kinzel, J. Kleinz, and R. Nehl, *J. Phys.* **A23**, L581 (1990).

93. G. Gyorgyi and N. Tishby, in *Neural Networks and Spin Glasses*, ed. W. K. Theumann and R. Koberle (World Scientific, 1990), p. 3.

94. H. Sompolinsky, N. Tishby, and H. S. Seung, preprint (1990).

95. G. Györgyi, *Phys. Rev.* **A41**, 7097 (1990).

96. H. Sompolinsky and N. Tishby, preprint (1990).

97. J. A. Fodor, *The Modularity of Mind* (MIT Press, 1983).

98. B. Julesz, *Rev. Mod. Phys.*, to appear.

99. I. Rock and S. Palmer, *Sci. Am.* **263**, 48 (1990).

100. M. Abeles, E. Vaadia, and H. Bergman, *Network* **1**, 13 (1990).

101. D. J. Amit and A. Treves, *Proc. Natl. Acad. Sci. (USA)* **86**, 465 (1990).

102. N. Rubin and H. Sompolinsky, *Europhys. Lett.* **10**, 465 (1989).

103. C. von der Malsburg and W. Schneider, *Biol. Cybern.* **54**, 29 (1986).

104. C. von der Malsburg and E. Bienenstock, *Europhys. Lett.* **3**, 1243 (1987).

105. R. Eckhorn, R. Bauer, W. Jordan, M. Brosch, W. Kruse, M. Munk, and R. J. Reitboeck, *Biol. Cybern.* **60**, 121 (1989).

106. C. M. Gray, P. König, A. K. Engel, and W. Singer, *Nature* **338**, 334 (1989).

107. H. Sompolinsky, D. Golomb, and D. Kleinfeld, *Proc. Natl. Acad. Sci. (USA)* **87**, 7200 (1990).

108. L. F. Abbott, *Proc. Natl. Acad. Sci. (USA)*, in press.

109. Y. Kuramoto, *Chemical Oscillations, Waves, and Turbulence* (Springer, 1984).

110. J. Buhmann, *Phys. Rev.* **A40**, 4145 (1989).

111. D. J. Amit, M. Evans, and M. Abeles, *Network* **1** (1990).

112. F. H. C. Crick and C. Koch, *Semin. Neurosci.* **2**, 263 (1990).

The Origins of Order
Self-Organization and Selection in Evolution

Stuart A. Kauffman

Department of Biochemistry and Biophysics, University of Pennsylvania
Philadelphia, PA 19104, USA

Introduction

In this chapter I will discuss the relation between selection and self-organization. I begin these efforts by discussing the structure of the rugged fitness landscapes underlying evolution. I shall consider such landscapes as fixed in structure. In reality, fitness landscapes deform due to changes in the abiotic environment, and due to coevolution. In coevolutionary processes, the fitness of one organism or species depends upon the characteristics of the other species with which it interacts while all simultaneously adapt and change. A critical difference between evolution on a fixed landscape and coevolution is that the former can be roughly characterized as if it were an adaptive search on a "potential surface" or "fitness surface" whose peaks are the positions sought. In a coevolutionary process, there may typically be no such potential surface. The process is far more complex and interesting; however, in this chapter I confine myself to the fixed landscape. (For a discussion of coevolution, see my book, *Origins of Order*.)

As Francois Jacob pointed out in his 1977 essay "Evolution and Tinkering", (see also Jacob 1983) adaptation typically progresses through small changes involving a *local* search procedure in the space of possibilities. The paradigm is one of local "hill climbing" via fitter mutants towards some local or global optimum. Despite this transparent metaphor, such a process involves complex combinatorial optimization. In such optimization searches many parts and processes must become coordinated to achieve some measure of overall success, but conflicting "design constraints" limit the results achieved. One purpose of this chapter is to investigate how the extent of conflicting constraints alters the character of the "hills" to be climbed. We will find that increasing levels of conflicting constraints make the landscapes

61

more rugged and multipeaked.

The "hill climbing" framework is hardly new, for I borrow it with minor modifications directly from Sewell Wright (1931, 1932), one of the three major founders of population genetics. Wright introduced the concept of a space of possible genotypes. In one version of his idea, each genotype has a "fitness", and the distribution of these fitness values over the space of genotypes constitutes a *fitness landscape*. (Often Wright thought of the fitness of a given gene or genotype as a function of its *frequency* in the population (Wright *ibid.*, Provine 1986). I will use the simpler idea that each genotype itself can be assigned a fitness in this chapter.)

Depending upon the distribution of the fitness values, the fitness landscape can be more or less mountainous. It may have many peaks of high fitness flanked by steep ridges and precipitous cliffs falling to profound valleys of very low fitness. Or the fitness landscape, like the gentle Normandy countryside, may be smoothly rolling with low hills and gentle valleys.

In this framework, adaptive evolution in a population is a hill climbing process. The population can be thought of as a tight or loose cluster of "individuals" located at different points in the landscape. Mutations "move" an individual, or its offspring, to neighboring points in the space representing neighboring genotypes. Selection is reflected in differential reproduction by individuals with different fitness. Therefore, over time the cluster of individuals representing the population will "flow" over the fitness landscape. In the simplest cases, the population will climb to and cluster about one of perhaps a very large number of different *fitness peaks* (Ewens 1979, Crow and Kimura 1965, 1970, Gillespie 1983, 1984). In more complex cases, the cluster representing the population may spread widely across the landscape, passing via a rich web of ridges somewhat below the fitness peaks (Eigen 1985, Schuster 1986, 1987, Fontana and Schuster 1987, Kauffman and Levin 1987, Kauffman *et al.* 1988, Kauffman 1989a, Eigen *et al.* 1989). In other cases the population may drift down from the peaks and wander within a band of modest fitness "altitudes" virtually anywhere across the fitness landscape (Fontana and Schuster 1988, Kauffman *et al.* 1988, Kauffman 1989a, Eigen *et al.* 1989).

It is intuitive from this description that the behavior of an adapting population depends upon how mountainous the fitness landscape is, on how large the population is, and on the mutation rate which "moves" an individual from one genotype to another genotype in the space. The flow of a population over a fitness landscapes also depends upon whether it is sexual, where mating allows mixing of genotypes from distant points in the landscape in a new individual, or asexual. In this chapter I will develop a body of theory to characterize the *mountainous structure* of multipeaked fitness landscapes. In particular, I introduce a general and flexible model, called the *NK model*, which is among the simplest possible models of "complex" systems adapting on mountainous "fitness landscapes". This model can be "tuned" to produce landscapes ranging from smooth hills to extremely jagged multipeaked moonscapes. By examining the statistical structure of such landscapes as their "ruggedness" is tuned we will be led to characterize many central features of adaptive evolution.

In *Origins of Order* I examine the biological implications of rugged fitness landscapes and the actual flow of an adapting population over such rugged landscapes under the drive of mutation, recombination and selection. This is, of course, the province of population genetics. An enormous amount of work has been done in the past half a century, far too vast to summarize here (Ewens 1979, Crow and Kimura 1965, 1970). Nevertheless, no adequate theory linking such population flow with the structure of very complex rugged landscapes yet exists.

In the present chapter it will prove convenient to conceive of a space or *ensemble* of possible "systems" each "next to" slightly different versions of the system. In the cases we will examine, most but not all members of the ensemble exhibit some spontaneous "ordered" properties. The ordered property may or may not have anything to do with fitness. If the ordered property is itself selected, we might of course expect to see the property in organisms. But if the ordered property is *not* under selection, or even modestly selected against, might we still see it? The answer can be "yes". The conditions when this can happen are the central issue.

The point here is similar to imagining an ensemble of objects most of which are blue, but which have a variety of other properties of size, weight, and so forth. If selection acts on size and weight, will we see blue objects in the continuing presence of selection?

There are two ways this might happen.

1) If under the conditions of adaptation the population does *not* remain tightly clustered over single mountain peaks of high fitness, but wanders within some larger volume of the ensemble, the chances are high that most members of that volume exhibit the ordered property. This corresponds directly to an analogy of selection as a *weak Maxwell's Demon*. If the

Demon operating the flap valve between the adjacent boxes to partition the faster molecules into the right box opens and closes the flap value rather slowly compared to the velocity of the molecules, then the statistical distribution of the faster and slower gas molecules in the two boxes will not differ sharply from thermodynamic equilibrium. Similarly if selection is too weak to hold an adapting population in very small volumes of the ensemble, then even in the presence of continuing selection the adapting population will almost certainly exhibit the "typical" ordered properties of most ensemble members. Hence I tend to use the phrase that such adapting systems would exhibit order not because of selection, but despite it.

2) The second fundamental way that adapting populations might be expected to continue to exhibit the spontaneous order typical of most members of the ensemble, even *if selection can hold the population within very small volumes* of the ensemble representing adaptive peaks, is that the vast majority of *the adaptive peaks themselves remain typical of the ensemble as a whole*. This limitation tends to become powerful as landscapes become very rugged and multipeaked. In such landscapes, the adaptive process tends to become "trapped" on local peaks and thus cannot move long distances to rare regions of the space. If there are many peaks, most of them are likely to be in typical regions of the space. Therefore, if the adaptive process starts at some "typical" spot in the space of possibilities, it will become trapped on a local peak which is very likely to remain typical of the space as a whole. "Blue" organisms will be found. But a related further feature of very rugged landscapes will soon emerge: As many rugged landscapes become more multipeaked, the fitness peaks themselves *dwindle in altitude* to mere hills, then hummocks, then faint bumps. As the peaks fall ever lower, they necessarily become progressively more typical of the space as a whole. Thus even when selection is very powerful and can hold populations on any accessible peak, the peak itself almost certainly exhibits properties typical of the entire space of possibilities. Thus if the space is the space of genetic regulatory systems and member systems typically exhibit the property that the number of cell types is about the square root of the number of genes, that relation can be expected across phyla despite strong selection. In the simple image: "Blue" organisms will be found.

These two limitations constitute what I shall call *two complexity "catastrophes"*, for we shall see that one or the other must ultimately occur as the *complexity of the entities under selection increases.* This is the most important point of the theory discussed in this chapter. As the complexity of entities increases, one or the other basic mechanism ultimately *limits the power of selection.*

The intuitive reasons for these limitations are not hard to see, and the *NK* model I introduce in this chapter makes the case clearly. In an adapting system of many parts, either those parts are fully independent of one another, or coupled together. In the limiting case where the parts are independent, each part typically makes a contribution to the overall function of the system which *decreases* in relative importance as the total number of parts in the system increases. For a system with a sufficient number of parts, the fitness loss due to mutational damage of one part becomes small. Therefore the selective force tending to restore the damage becomes weaker than the mutational "pressure" tending to damage the part. In short, selection becomes too weak a force to hold an adapting population at adaptive peaks. The population flows down the adaptive hillside to the lowlands. This contention of mutational and selective forces leads to a complexity or "error" catastrophe when the number of parts exceeds a critical value. Beyond that complexity, selection cannot climb to peaks or remain here.

At the opposite extreme the parts are very richly coupled. But in this case common experience suggests that conflicting "design constraints" will make it difficult to achieve overall success. As we will soon see, such conflicting constraints lead to an adaptive landscape which becomes more multipeaked as the number of parts increases. Thus adaptation, which must search such rugged landscapes, tends to become trapped in very small regions of the space. Worse, due to the increasing numbers of conflicting constraints, the actual fitness peaks become ever poorer compromises among those constraints. The peaks wither to mere bumps hardly better than chance agglomerations of the parts.

These investigations suggest that adaptive evolution is bounded by the character of fitness landscapes. But that character in turn depends upon the entities which are evolving. Hence evolution can change the rugged structure of fitness landscapes and their impact on evolution by changing the adapting entities themselves. Thus we will ask what kinds of landscapes, in what conditions, allow adaptive evolution to be optimized.

Fitness Landscapes in Sequence Space

Sequence Space: The "Practical" Importance of a Theory
of Adaptation on Rugged Landscapes

The framework just sketched has, as one overarching purpose, the aim
of analyzing the relation between self ordering and selection in complex
systems. But there are more immediate reasons to develop a theory of
adaptation on rugged "fitness landscapes". The same framework provides a
crisp means to describe the selective evolution of proteins or RNA molecules
for specific functions. More generally it applies to adaptive evolution in
"sequence spaces".

Before continuing, I must clarify what I shall mean by a fitness landscape. For an evolutionary biologist, "fitness" is a term which applies prin-
cipally to an entire organism. It has components of fecundity, fertility,
and other factors leading to reproductive success (Ewen 1979, Crow and
Kimura 1965, 1970). These include complex issues such as the *frequency*
of each genotype variant of the organism in the population, the *density* of
each genotype variant in a region, and even the entire *ecosystem* with which
each organism interacts (Levin 1978). Therefore, in the general context it
is difficult to assign a fitness to a gene, or even a genotype, since account
must be taken of those factors which depend upon other organisms.

For the purposes of the present chapter I shall use the term "fitness landscape" in a much more restricted sense to refer to any well defined property
and its distribution across an ensemble. For example, I will shortly define
"protein space" more precisely. The capacity of each protein in protein
space to catalyze a specific reaction under specified conditions is, in principle, also a well specified property. The velocity of the reaction catalyzed by
each protein can then be *defined* as the "fitness" of that protein. Then the
distribution of velocities across the space of proteins constitutes the "fitness landscape" with respect to that defined function. Adaptive evolution
with respect to that specific function is a search process in protein space
which attempts to optimize the capacity to catalyze that specific reaction.
It is an entirely different issue whether optimization of any specific reaction
velocity optimizes the overall fitness of the organism harboring the enzyme.

The concept of "protein space" was, to my knowledge, first introduced
by the evolutionary biologist J. Maynard Smith (1970). It has since been
reinvented by a number of authors including J. Ninio (1979), and more recently M. Eigen (1985, 1987), and Schuster (1986, 1987). Others, (Borstnick

et al. 1987), including myself and colleagues, (Kauffman and Levin 1987, Kauffman *et al.* 1988, Kauffman 1989a,b) have utilized Maynard Smith's initial idea as well. Proteins are linear polymers comprised of 20 different kinds of amino acids. Because proteins have distinguished carboxy and amino terminal ends (Alberts *et al.* 1983), each polymer is oriented. The total number of proteins of a specific length, N, is just 20^N. Therefore, this set of all possible proteins length N constitutes an *ensemble*. Furthermore, each protein can be mutated to other proteins by changing any amino acid at one position in the protein to one of the 19 other possible amino acids. Therefore, for a protein length N, there are $19N$ "1-mutant" neighbor proteins. The concept of a *protein space* is a high dimensional space in which each point represents one protein, and is next to $19N$ points representing all the 1-mutant neighbors of that protein. The protein space therefore simultaneously represents the entire ensemble of 20^N proteins and keeps track of which proteins are 1-mutant neighbors of each other.

There is a certain charm in these discrete spaces, for we can specify precisely what we mean by neighboring sequences, the minimum number of changes to pass from one sequence to another, and so forth. But the concept is in fact very general. Thus, in my book *Origins of Order* I use a node to represent, not a specific sequence, but also complex entities such as an entire geonomic cybernetic system. Again, the concept of neighboring nodes carries over to systems which are "one-mutant neighbors" from one another. Thus high dimensional discrete spaces are a convenient way to represent *ensembles* of related objects where it is natural to keep track simultaneously of which objects are 1-step neighbors of one another.

John Maynard Smith's purpose in invoking protein space was to note that if adaptive evolution in general occurs by substitution of single amino acids, then evolution is a "walk" between adjacent nodes in protein space. Therefore, to improve function, any such *adaptive walk* in general must be a *connected* walk through a succession of adjacent nodes all of which exhibit *improved function*.

There are a number of obvious questions to ask about such adaptive walks:

1) How many local optima exist in the space?
2) What is the average number of improvement steps on an adaptive walk to a local optimum (that is, a peptide sequence which is fitter than all its 1-mutant neighbors)?

3) What is the average number of mutants "tried" on an adaptive walk to a local optimum?

4) What is the ratio of the mutations "accepted" to those "tried"?

5) What is the average number of alternative local optima which can be reached from a peptide in the space? What is the maximum?

6) After each improvement step the number of fitter 1-mutant neighbors may change. At each local optimum there are 0 fitter 1-mutant neighbors. Therefore, we might expect the average number of fitter 1-mutant neighbors to *dwindle to 0* on an adaptive walk. How does it dwindle?

7) How many peptides can "climb to" the same local optimum?

In analyzing the *statistical structure* of fitness landscapes I will make use of the simplest image of an adaptive walk via fitter 1-mutant variants. This is an idealization of the actual flow of an adapting population which I adopt in this chapter in order to consider how mountainous such high dimensional landscapes might be expected to be. Nevertheless, it is important to stress that this idealized image of walks constrained to pass only via fitter 1-mutant neighbors corresponds to one plausible limiting case of the adaptive flow of a real population under the drives of mutation, selection, and recombination. Gillespie (1983, 1984) has shown that this version of an adaptive walk corresponds to a population adapting on a landscape such that the rate of finding fitter variants is very slow compared to the fitness differentials between the less fit and the more fit allele. In such a limit, if the population begins entirely at the less fit allele, a single mutant will eventually encounter the fitter allele. Either by chance that mutant dies out before leaving offspring, or a few of the fitter mutant type are produced. Once a sufficient number of the fitter type are produced to reduce the chance fluctuation leading to their death, the fitter type takes over the entire population on a rapid time scale. Thus the entire population "hops" to the fitter neighboring genotype. Gillespie has shown that the entire adaptive process in this limit can be treated as a continuous time, discrete state Markov process. Each state corresponds to one genotype. The population jumps as a whole with different probabilities to one or another of the fitter neighboring genotypes. The conditions required for Gillespie's limit are the product of population size and mutation rate is low compared to the rate of finding fitter variants.

The *NK* Model of Rugged Fitness Landscapes

The *NK* Model of Random Epistatic Interactions

I now introduce and discuss in detail a simple formal model of rugged fitness landscapes, called the *NK* model. In this model, *N* refers to the number of "parts" of a system, genes in a genotype, amino acid in a protein, or otherwise. Each part makes a fitness contribution which depends upon that part itself and upon *K* other parts among the *N*. That is, *K* reflects how richly crosscoupled the system is. In the geneticist's term defined more precisely below, *K* measures the richness of epistatic interactions among the components of the system.

Since the model is abstract, I should make it clear why I believe analysis of the model warrants detailed attention. The real mountainous structures of fitness landscapes with respect to catalytic or other functions in proteins are unknown, but knowable. Discovery of the actual structures of such landscapes is of the deepest importance. Although we do not yet know what the real landscapes are like, we may nevertheless be able to develop some intuition for their typical, or statistical structures by building simple models. That is, we need a kind of statistical mechanics for fitness landscapes helping us to understand their expected features. The *NK* model I now introduce is meant to accomplish this. The model generates a *family* of increasingly rugged multipeaked landscapes as the main parameters, *N*, *K*, and others described below, are altered. At the moment, this model is perhaps the simplest, with perhaps the best understood family of landscapes, available. Therefore, analysis of its features can only be useful.

A second reason to develop a formal model for the statistical structure of rugged fitness landscapes lies in the fact that we want to predict and understand the structure of actual fitness landscapes in protein space and elsewhere. The *NK* model is the first effort in this direction. Perhaps surprisingly, given its simplicity, the model performs rather well when confronted by known adaptive landscapes in protein space. For example I have elsewhere that it leads to a variety of realistic and testable predictions about adaptive somatic evolution in the immune response. (Kauffman 1991). The success attained by the model lends hope to the possibility that very good statistical models of landscape structure may ultimately be constructed.

The model can be interpreted in terms of a fitness landscape in protein space. However, the same model can be interpreted as a model of genetic interactions within a genetic framework. One of the earliest population genetic models focuses on haploid organisms with a single copy of each chromosome. Each chromosome has some number of distinct genes, and the chromosome set has a total of N distinct genes. Each gene may occur in more than one version, or allele. In the simplest case, each gene can occur in two different alleles (Ewens 1979, Crow and Kimura *ibid.*). Then the haploid genotype has N genetic loci, each with 2 alleles. More generally, each locus might have some larger number of alleles, A. In the first case, of course, the total number of genotypes is 2^N. In the latter case the total number of possible genotypes, if each locus has exactly A alleles, is A^N.

In keeping with the ensemble framework, the set of possible genotypes constitutes the ensemble. Each genotype is a 1-mutant neighbor of all those accessible by mutating a single locus from one to another allele. In the N-locus, 2-allele case each genotype is a 1-mutant neighbor of N other genotypes. In the A allele per locus case, each genotype is the 1-mutant neighbor of $N(A-1)$ other genotypes. The number of 1-mutant neighbors is the *dimensionality* ∂ of the genotype space giving the number of directions in which each genotype can change by a minimal alteration to another neighboring genotype.

The assumption that each gene contributes to overall fitness *independently* of all other genes, is clearly an idealization. In a genetic system with N genes, the fitness contribution of one or another allele of one gene may often depend upon the alleles of some of the remaining $N-1$ genes. Such dependencies are called *epistatic interactions*. The existence of epistatic interactions is well known. Their existence raises experimental and theoretical issues. The experimental issues concern how to *measure* the extent of epistatic interactions. I will not discuss this further. The theoretical problem is how to build *useful models* of epistatic interactions. One device commonly used has been to assume that genetic loci which interact can be represented by *multiplying* their fitness contributions (Franklin and Lewontin 1970, Ewens 1979, Lewontin 1974). Multiplication captures a kind of positive cooperativity. High fitness contribution by two epistatic loci requires that *both* have high fitness. If either has low fitness, then the product may be low even if the other has high fitness.

The problem with any such model is that the ways in which different alleles at the N loci might be coupled to one another epistatically to produce an overall fitness to each genotype might be extraordinarily complex. In general we truly have almost no idea what those mutual influences on overall fitness might be. Take Mendel's own peas. He found two alleles for color of seed, yellow and green, and two alleles of a second gene for texture of seed, rough and smooth. A *priori* we have no idea which of the four combinations of these traits will be of highest fitness, nor how changing from any one combination of traits to any other will affect fitness. If the fitness contribution of each gene is epistatically affected by a large number of other genes, the possible *conflicting constraints* among the complex web of epistatically interacting genes are both unknown, and likely to be extremely complex. This suggests that it might be useful to confess our total ignorance, and admit that for different genes, and those which epistatically affect them essentially arbitrary interactions are possible. Then we might attempt to capture the *statistical features* of such complex webs of epistatic interactions by assuming that the epistatic interactions are so complex that we can model the *statistical features* of their consequences with a *random fitness function*, this leads to the NK model.

Consider an organism with N loci, each with two alleles, 1 and 0. Let K stand for the average number of other loci which epistatically affect the fitness contribution of each locus. Thus the two main parameters of the NK model are the number of genes, N, and the average number of other genes, K, which epistatically influence the fitness contribution of each gene. The third parameter characterizes how the K genes which epistatically affect each gene are distributed among the N genes. For example, the genes are located in a spatial order along one or more chromosomes. One might want to suppose that the K genes which influence each gene are its *spatial neighbors* along the chromosome. Alternatively, the K genes influencing each gene might be chosen at random from among the N genes, or in some non-random way which is a function of distance of the loci from one another. In addition, one might want to suppose a *reciprocity*. If gene I influences gene J, then J influences I. In general the sensible steps to take with such models is to assess which parameters actually matter. It will turn out that to a very large extent *only N and K* matter. The distribution of K among the N appears to be far less important. A fourth parameter, A, is the number of alleles at each genetic locus. For the moment restrict this to two.

Having assigned to each locus, i, the K genes which impinge upon it, it is necessary to assign fitness contributions to each gene in the context of the K which epistatically influence it. The fitness contribution of the allele at the ith locus depends upon itself, 1 or 0, and the alleles, 1 or 0, at K other loci; hence, upon $K + 1$ alleles. The number of combinations of these alleles is just $2^{(K+1)}$. Since we have no idea what the effects of each such combination on the fitness contribution at the ith locus might be, let us *model* those effects by assigning to each of the $2^{(K+1)}$ combinations, *at random, a different* fitness contribution drawn from the *uniform distribution between 0.0 and 1.0*. Therefore, the fitness contribution, w_i, of the ith locus is specified by a *list, or vector of random* decimals between 0.0 and 1.0, with $2^{(K+1)}$ entries.

The fitness contribution of each of the alleles at *each of the N genes*, in the context of the K other genes which impinge upon that gene, must be specified. For each gene its fitness contribution, w_i, is generated by random assignment on the $2^{(K+1)}$ allele combinations of the $K+1$ genes which impinge upon it.

Having assigned the fitness contributions of each allele of each gene in the context of the alleles at that gene, and, the K others which epistatically act on that gene, we may now define the *fitness of the entire genotype* as the *average* of the contributions of each locus. That is, add the fitness contributions of the N alleles and divide by N:

$$w = \frac{1}{N} \sum_{j=1}^{N} w_i .$$

(1)

Given this definition of the fitness of each entire genotype, the NK model is fully specified. Therefore, the fitness of each possible genotype in the space has been assigned, and in consequence, a fitness landscape over the genotype space has been created.

The NK model, as defined, is based on only 2 alleles at each of the N genetic loci. Extension to any arbitrary number of alleles, A, at each locus is obvious. For each genetic locus, i, with epistatic interactions from K other loci, its fitness contribution must be specified for all $A(K+1)$ combinations of alleles present at the $K+1$ loci. As before, these fitness contributions are assigned at random from the uniform interval between 0.0 and 1.0.

This completes the definition of the NK model. It is disarmingly simple. The model merely says that each gene makes a fitness contribution which depends upon itself and K other genes in ways which are so complex that the epistatic interactions are to be modeled by a random fitness function. Nevertheless it is flexible, general, and surprisingly powerful. In particular, the existence of complex webs of epistatic interactions captures the fact of conflicting constraints. Such constraints would be expected to alter the mountainous character of the fitness landscape. Indeed, we are about to see that tuning K from 0 to $N-1$, thereby increasing epistatic interaction, tunes the corresponding fitness landscape over the space of genotypes from the very smooth to extremely rugged.

I have introduced the NK model in terms of a genetic model with epistatic interaction among N genes. But notice that the "site may equally be interpreted as positions in the primary sequence of a protein, and $A = 20$ alleles interpreted instead of the $A = 20$ amino acids which might occur at each position. Therefore, the NK model can also be thought of as a model of fitness landscapes in protein space. Indeed, the N sites can be considered as N "traits" in an organism, present or absent, and the NK model can be interpreted as a measure of conflicting constraints arising due to epistatic interactions among traits in a whole organism. I have utilized this in thinking about the Cambrian explosion of phyla and branching phylogenies (Kauffman 1991). For consistency, in the remainder of this chapter I will use the genetic interpretation.

The NK model is very similar to *spin-glass* models (Edwards and Anderson 1975, Sherrington and Kirkpatrick 1975, Anderson 1985, Binder and Young 1986). Indeed, Anderson has pointed out that the NK model is a form of spin glass. This is important to us for two reasons. First, a feature of spin-glasses called *frustration* helps account for the multipeaked features of fitness landscapes. Second, there are profound similarities between the behaviors of a physical system in a complex potential surface at a finite temperature and an adapting population on a rugged fitness landscape at a finite mutation rate. The tools of statistical physics bear on population biology.

Conflicting constraints in spin-glasses account for the rugged structure of their potential surfaces. Consider a set of four adjacent spins in a square lattice spin glasses where three of the four pairs prefer to point in the *same direction*, while the fourth prefers to point in the opposite directions. No arrangement of spin up and spin down states of the four spins around the square can satisfy all these constraints. Such a square is said to be *frustrated* (Anderson 1985). This frustration, due to conflicting constraints, leads to a complex energy surface with very many potential minima. We will find the direct analogue in the NK model, for as K increases the conflicting constraints leads to an ever more rugged multipeaked fitness landscape.

As already stated, a large number of properties of the landscapes created by the NK model appear to be surprisingly robust, and depend almost exclusively upon N and K alone. Therefore it is important at the outset to explain which features are dependent on other aspects of the model.

One very sensitive feature of the model is the actual *range of fitness* assigned to the space of genotypes. This depends upon the assumption that the fitness value assigned to each of the $A^{(K+1)}$ combinations of alleles bearing on each allele, i, was drawn at random from the *uniform interval* between 0.0 and 1.0. I might instead have assigned values at random from a different underlying distribution, for example a peaked Gaussian distribution between 0.0 and 1.0 in which the random decimals are more likely to be near 0.5 than near 0.0 and 1.0, or a U-shaped distribution between 0.0 and 1.0 in which the random decimals are more likely to be near 1.0 or 0.0 than near 0.5. In the Gaussian case, this would tend to "squeeze" fitness values assigned to all possible genotypes closer to the mean of that distribution, 0.5. Use of the U-shaped distribution would tend to expand the derivation of fitness values assigned to all possible genotypes further away from the mean fitness of the ensemble, 0.5. Since the actual fitness values assigned are sensitive to the choice of the underlying distribution used, I shall avoid properties of the NK model which are known to be sensitive to this choice.

The reasonably insensitive properties of the fitness landscapes generated by the NK model appear to include:

1) The number of fitness peaks in the genotype space.
2) The lengths of walks via fitter neighbors to fitness optima. Equivalently, this is the number of "accepted" mutations on an adaptive walk.
3) The time or total number of mutations "tried" before reaching an optimum.
4) The ratio of accepted to tried mutations on an adaptive walk.
5) The number of alternative optima to which one genotype can climb.
6) The number of genotypes which can climb to the same optimum.
7) The rate at which the fraction of fitter neighbors *dwindles* to 0 along adaptive walks to fitness peaks.
8) The similarity of local optima.

As N and K are changed, these statistical features of the corresponding more or less jagged fitness landscapes over genotype space alter. These statistics, however, are largely insensitive to the actual range of fitness values due to the choice of underlying uniform, Gaussian, U-shaped or other distributions. This insensitivity rests on the fact that, for the moment, I will count a neighboring genotype "fitter" than another even if the fitness difference is infinitesimal. Ignoring the actual fitness differences amounts to a kind of rank ordering of the fitness of all possible genotypes. In this case, adaptive walks will pass from any genotype to any fitter genotype.

I should emphasize that these properties can, at present, only be said to appear to be insensitive to the underlying distributions from which fitness values are assigned. This assertion is based on numerical investigations (Kauffman *et al.* 1988). The extent of insensitivity warrants further study.

By focusing on the rank order statistics of the NK family of landscapes we achieve a class of models which yields substantial insight into the statistical structure of rugged fitness landscapes. However, the actual *flow* of an adapting population on such a landscape will also depend critically on the actual fitness differences between adjacent genotypes in the space.

We turn next to an examination of landscape structure as a function of the main parameters of the model, N and K. I discuss first the two extremes, $K = 0$, which corresponds to the limit of a smooth landscape with a single fitness peak, and $K = N - 1$, which corresponds to a completely random fitness landscape with very many fitness peaks. Thereafter I characterize the family of correlated landscapes which lies between these extremes.

The first case to examine is the $K = 0$ limit of the NK model, in the further condition that each gene has only two alleles. Then there are no epistatic interactions. Each locus makes a contribution to fitness independently of all other loci. From Eq. (1), the fitness of any genotype with a defined combination of alleles at the N distinct loci is just the sum of the contributions of the independent loci divided by N. This is exactly the N locus two allele additive fitness model for a haploid system (Ewens 1979, Crow and Kimura 1965, 1970).

We show next that the *structure* of this fitness landscape has a single global optima genotype, all other genotypes are suboptima and can climb to the global optimum via fitter neighbors, and all 1-mutant neighbors have nearly the same fitness.

Let 0 and 1 stand for the two possible alleles at each locus. At each locus, by chance, either allele 0 or allele 1 makes the higher fitness contribution. Therefore, there is a special genotype with the fitter allele at each of the N loci which is *the global optimum genotype*. Furthermore, any other genotype, which must of course have lower fitness, can be sequentially changed to the globally optima genotype by successively "flipping" each gene which is in the less favored allele to the more favored allele. Therefore, any such suboptimal genotype *lies on a connected pathway via fitter 1-mutant variants to the global optimum.* It follows trivially that there are no optima other than the single global optimum. All other genotypes are both below the global optimum and can climb to it.

In previous sections I have used the pictorial image of a "rugged" or "smooth" fitness landscape without defining the term. A more precise term is the "correlation structure" of the fitness landscape. By this I mean how similar the fitness values of 1-mutant neighbors in the space are. A *smooth* landscape is one in which neighboring points in the space have nearly the same fitness value. Knowing the fitness value of one point carries a lot of information about the fitness value of neighboring points. At the opposite extreme, the fitness values might be entirely *uncorrelated*. Knowing the fitness at one point would then carry no information about the fitness of neighboring points. A variety of alternative measures can be used to characterize the correlation structure of a fitness landscape.

The $K = 0$ additive model corresponds to a very smooth, highly correlated fitness landscape. This is clear because the fitness of 1-mutant neighbors cannot differ by more than $1/N$. Therefore, for large N, the fitness of 1-mutant neighbors is very similar.

Two other features of the $K = 0$ model with 2 alleles per locus are immediately understandable. If an adaptive walk starts anywhere and climbs via successive fitter 1-mutant variants, then the number of fitter neighbors dwindles by 1 at each step upward. If the walk starts with a randomly chosen genotype, on average half the N loci are already in the more favored allele, and the other half are in the less favored allele. Therefore, the expected number of steps to the optimum is just $N/2$. This implies that walk lengths to the global optimum increase *linearly* as N increases.

$K = N - 1$

The largest possible value of K is $N - 1$. In this limit, each gene is epistatically affected by all the remaining genes. It is particularly easy to show that in this limit the resulting fitness landscape is entirely uncorrelated. The fitness value of one genotype gives no information about the fitness value of its 1-mutant neighbors. As I show in the next section, we can also understand a number of quite surprising features of such extremely rugged fitness landscapes (Kauffman and Levin 1987, Weinberger 1988, Macken and Perelson 1989). In particular, we will see that:

1) The number of local fitness optima is extremely large.
2) The expected fraction of fitter 1-mutant variants dwindles by 1/2 at each improvement step.
3) The lengths of adaptive walks to optima are very short, and increase only as a logarithmic function of N.
4) The expected time, or number of mutants "tried", to reach an optimum is proportional to the dimensionality of the space.
5) The ratio of accepted to tried mutations scales as $\ln N/N$ for two allele case.
6) Any genotype can climb to only a small fraction of the local optima.
7) Only a small fraction of the genotypes can climb to any given optimum.

The 7 features above are ordering properties of completely uncorrelated landscapes. Perhaps the most important implication of these fully uncorrelated landscapes, which as we shall soon see carries over to a large class of rugged but correlated landscapes, is this:

8) As the number of genetic loci, N, increases the local optima fall toward the mean fitness of the space of genotypes.

The eighth implication is so central that I shall call it a further kind of *complexity catastrophe*. It points to a fundamental restraint on adaptive selection. Conflicting constraints in complex systems limit the optimization of function which is possible. As we shall see, this appears to be a very general problem for many classes of systems.

The $K = N - 1$ limit corresponds to completely uncorrelated fitness landscapes. The fitness vector w_i for each gene i, $i = 1, 2, \ldots, N$, is a function of all $K + 1 = N$ genes. Consider any initial genotype among the 2^N genotypes with two alternative alleles at each locus. Alteration of the allele at any single locus affects each of the N genes, since it alters the combination of the $K + 1 = N$ alleles which bear on the fitness of each gene. In turn, this alters the fitness contribution of each gene to a different randomly chosen value between 0.0 and 1.0. The fitness of the new 1-mutant neighbor genotype is therefore a new sum of N random decimals between 0.0 and 1.0. Therefore the new fitness value is *entirely uncorrelated* with the old fitness value. Since fitness values of 1-mutant neighbors are entirely random with respect to one another, the $K = N - 1$ landscape is fully uncorrelated.

The first point to stress is straightforward. $K = 0$ corresponds to fully correlated smooth landscapes. $K = N - 1$ corresponds to fully uncorrelated rugged random landscapes. Therefore as K increases, landscapes must change from smooth through a family of increasingly rugged landscapes. *Increasing the richness of epistatic interactions, K, increases the ruggedness of fitness landscapes.* Since increasing epistatic interactions simultaneously increases the number of conflicting constraints, increased multipeaked ruggedness of the fitness landscape as K increases reflects those increasingly complex mutual constraints.

The Rank Order Statistics on $K = N - 1$ Random Landscapes

1. *The number of local optima is very large*

We now calculate the expected total number of local optima with respect to 1-mutant neighbors. In keeping with the hypothesis that walks must pass via fitter 1-mutant neighbors, regardless of how small the fitness differentials may be, it is convenient to rank order all the genotypes from worst (1) to best (2^N). The probability, P_m, that any genotype is a local optimum is just the probability that it has higher rank order than any of its N 1-mutant neighbors:

$$P_m = 1/(N+1). \qquad (2)$$

Since the total number of genotypes with 2 alleles per locus is 2^N, the expected total number of local optima with respect to 1-mutant moves, M1, is

$$M1 = (2^N)/(N+1). \qquad (3)$$

Therefore, the number of local optima is extremely large and increases almost as rapidly as the number of genotypes, 2^N. This means that these extremely rugged landscapes are so rife with local optima that trapping on such optima is essentially inevitable.

While I have considered only walks via fitter 1-mutant neighbors, it is useful to calculate the number of local optima if walks can proceed by 2-mutant, or R-mutant neighbors. The denominator in Eq. (3) is replaced by the total number of genotypes which can be reached in R or fewer mutations. This is just the cumulative binomial sum

$$\sum_{j=0}^{R} \binom{N}{j} \tag{4}$$

where $R = 1$ in Eq. (3). Thus, for any small value of R, as N increases the number of genotypes increases exponentially, but the number of local optima with respect to walks via fitter R-mutant neighbors increases very rapidly as well.

It is easy to generalize the $K = N = 1$ model from 2 alleles per gene to an arbitrary number of alleles per locus, A. The number of genotypes in the space is then A^N. As defined earlier, the *dimensionality*, D, of the space, (or equivalently the number of 1-mutant neighbors to each genotype, D) is $N(A - 1)$. Substitution into Eq. (3) by the number of genotypes in the numerator, and the number of 1-mutant neighbors in the denominator, gives the expected number of local optima:

$$(A^N)/(D+1). \tag{5}$$

2. *The expected fraction of fitter 1-mutant neighbors dwindles by 1/2 on each improvement step*

It is also clear that, at each improvement step, the expected fraction of fitter 1-mutant neighbors dwindles by 1/2. The landscape is entirely uncorrelated. Let the adaptive walk begin from the lowest rank genotype. All its D neighbors are fitter, with rank orders spread randomly between 2 and A^N. The adaptive walk samples neighbors at random and moves to the first fitter one encountered. Since those fitter neighbors are spread uniformly in rank order from just above the current genotype to the top, and a random fitter neighbor is picked, on average its rank order lies *halfway to the top*. When the process moves to that neighbor, because it is expected to be half way to the top, only half its 1-mutant neighbors are still fitter. On average, each successive step jumps again half the remaining distance to the top rank, hence at each step the expected number of fitter 1-mutant neighbors dwindles by 1/2.

This argument replaces the mean of a family of such adaptive walks with a "mean walk". In short, on random landscapes the number of "ways" uphill decreases rapidly. Recall, by contrast, that in $K = 0$ smooth landscapes, the number of ways uphill decreases only by 1 at each improvement step.

3. *Walks to local optima are short and vary as a logarithmic function of N*

In general, adaptive walks might begin anywhere. However, to obtain an upper bound on walk lengths we consider walks which begin at the lowest ranked genotype. An adaptive walk which begins at the lowest ranked genotype and steps halfway to the top of each step yield an expected *relative rank order* X/T, where T is the top rank:

$$X/T = (2^R - 1)/2^R \tag{6}$$

at each step R.

When the adaptive walk arrives at a genotype of relative rank order X/T, that genotype is fitter than at least the one from which the process left. Therefore the probability, P_m, that the newly encountered genotype is itself a local optimum is

$$P_M = (X/T)^{D-1}, \tag{7a}$$

When X and T are small this equation must be modified slightly to take account of the lack of replacement in calculating P_m (Kauffman and Levin 1987):

$$p_m = (X - 1)!(T - D - 1)/[(T - 1)\,(X - D - 1)!]. \tag{7b}$$

Combining Eqs. (6) and (7a) allows us to calculate the probability P_L that an adaptive walk continues for L steps *without* encountering a local optimum:

$$P_L = \prod_{R=0}^{i} \left[1 - \left(\frac{2^R - 1}{2^R} \right)^{D-1} \right]. \tag{8}$$

Each term in this product is one minus the probability that the current genotype reached with a relative rank order given by Eq. (4) is actually a local optimum, as given by Eq. (7a). Hence each term is the probability that the current genotype is not a local optimum, and therefore that the adaptive walk continues at least one more adaptive step.

As L increases, this product decreases and eventually falls below 0.5. The value of L at which this occurs is the number of steps taken such that, in half the trials, the walks will have arrested, while the rest may continue. Thus, Eq. (8) yields an estimate of the expected lengths of walks before a local maximum arrests progress upwards.

Equation (8) implies that adaptive walks in uncorrelated landscapes are surprisingly short, and tightly bounded. Note that if $R = \log_2(D - 1)$, the corresponding term in P_L is $1 - (1 - (1/(D - 1)^{(D-1)})$, which is extremely well approximated by $1 - 1/e = .63$. Moreover, if this represents the Rth term in P_L, it is easily shown that the preceding terms are approximately $1 - 1/e^2 = .86$ and $1 - 1/e^4 = .98$. Thus there is very little probability of the process stopping more than one or two adaptive steps before this value of R.

To a high degree of accuracy, the adaptive walk will stop on average at the Rth step when

$$R = \log_2(D - 1).\qquad(9a)$$

Equation (9a) therefore shows that the expected lengths of adaptive walks, R, in uncorrelated landscapes are *short*, on the order of the *logarithm base 2* of the number of neighbors to each entity in the space.

Weinberger (1988) and Macken and Perelson (1989) have carried out more detailed analysis of such walks, examining the entire distribution of walk lengths with similar results. Because adaptive moves which happen to step further than halfway to the "top" at each improvement are more likely to truncate an adaptive walk than adaptive moves which step less than half to the top are likely to lengthen the walk, Eq. (9a) is an overestimate of actual walk lengths to optima. Accounting for such fluctuations shows that adaptive walk lengths, R, are more nearly the natural logarithm of the number of 1-mutant neighbors. This feature shows that in highly rugged landscapes, there are so many mountain peaks that the local peaks which can be climbed from any point are very close in sequence space:

$$R \sim \ln(D - 1).\qquad(9b)$$

4. *The expected time to reach an optimum is proportional to the dimensionality of the space*

Consider a walk which begins at the lowest ranked genotype and climbs to a local optimum. Let the adaptive walk examine the D 1-mutant neighbors sequentially, taking one unit of time for each examination. This corresponds to the waiting time for a mutational event, but ignores the fact that such mutations do not examine all neighbors in order. The modification makes only a minor difference.

Since on average the adaptive walk steps halfway to the top rank at each improvement step, the *expected waiting time* to find a fitter variant *doubles* after each improvement step. This result is equivalent to the "theory of records" found in Feller's classical probability text (Feller 1971, volume 2). The first improvement step occurs after one moment, the second on average requires 2 moments and the third on average requires 4 moments. The expected number of improvement steps to each a local optimum is $\log_2(D-1)$. Thus the expected waiting time, T_{op}, to reach that optimum is just

$$T_{op} = \sum_{L=0}^{(\log_2 D-1)-1} 2^L \tag{10}$$

When $\log_2(D-1)$ is an integer, this series sums to $D-1$. Therefore the expected time to reach an optimum is proportional to the number of 1-mutant neighbors (or equally, the dimensionality of the space). The "time" to reach an optimum, of course, is also equivalent to the total number of mutants "tried" before reaching an optimum.

Macken and Perelson (1989) have redrived these results, and found a surprising additional fact: For large N and K, the time, or number of mutants "tried" before an optimum is reached, is nearly independent of the starting fitness.

5. *The ratio of accepted to tried mutations scales as lnN/N*

Since the length of an adaptive walk is the number of accepted mutations, while the time is number of "tried" mutations, the ratio of these is just $\ln N/N$ for the two allele case. Macken and Perelson (1989) make the same point with the addendum that these results may be quite insensitive to starting fitness.

6. *Any genotype can only climb to a small fraction of the local optima*

A maximal estimate of the number of branches to fitter variants which might emerge from the lowest ranked genotype can be obtained. D of its neighbors are fitter. On average, after a single improvement step, $(D-1)/2$, or almost $D/2$, of the neighbors of that first step variant are still fitter. After successive steps, on average $D/4, D/8, \ldots$ neighbors are fitter. Adaptive walks continue for about $\log_2(D-1)$ steps. The series $(D \times D/2 \times D/4 \times \ldots D/D)$ yields a gross upper bound on the expected number of alternative local optima accessible from the lowest ranked entity. This bound is

$$D^{(\log_2 D)}/2^{[\log_2 D][\log_2 D-1]/2} = D^{(\log_2 D)}/D^{(\log_2 D+1)/2}$$

$$= D^{(\log_2 D-1)/2}. \tag{11}$$

This bound is an overestimate since it ignores *convergence* of walks. Nevertheless, it suffices to establish the major point. Branching adaptive walks in uncorrelated landscapes reach only a small fraction of the total number of local optima. Thus, for a genotype space using $A = 20$ alleles per locus, and $N = 64$ genetic loci, an upper bound on the number of accessible local optima from the lowest ranked genotype is about 10^{14}; by contrast, the total number of local optima in such an uncorrelated genotype space is about 10^{80}.

This property means that any beginning sequence, no matter how poor, can only climb to a small range of local peaks via 1-mutant neighbors, and is limited to their fitnesses. Like New England, in rugged landscapes "You can't get there from here".

7. *A small fraction of the genotypes can climb to any one optimum*

It is useful to calculate an upper bound on this number. First, note that in general there is only a single genotype which is the global optimum is the landscape. We ignore "ties". Note next that the set of genotypes which can climb to that optimum via 1-mutant fitter neighbors is identical to the set of genotypes that the global optimum could *descend to* via 1-mutant *less fit neighbors*. Then the fittest genotype could reach D less fit 1-mutant neighbors. Each of those on average has $D/2$ still less fit neighbors and each of those has on average $D/4$ less fit neighbors. By the now familiar argument, this process continues for about $\log_2 D$ steps. Therefore an upper bound on the total number of genotypes which can climb to the global optimum is given by the sum $\{D + D \times D/2 + D \times D/2 \times D/4 + D \times D/2 \times D/4 \times D/8 \ldots D \times D/2 \times D/4 \times D/D\}$.

The formula is exact only if D is a power of 2, and is an upper bound because it ignores the possibility that a genotype might climb by two or more routes to the global optimum.

This series leads to an expression for the number of genotypes which can climb to the global optimum

$$\sum_{I=1}^{\log_2 D} D^I \cdot 2^{-[I(I-1)/2]} \tag{12a}$$

Consider genotypes with $N = 256$ loci and $A = 2$ alleles. The number of genotypes is 2^{256} or about 10^{77}, while the number of these which can climb to the globally optima genotype is only about 2^{29} or 10^9. Thus, only a tiny fraction of the genotypes can climb to even the global optimum if constrained to pass via fitter 1-mutant variants. This implies that the vast majority of adaptive walks via 1-mutant fitter variants end on optima which are below the global optimum.

I am indebted to D. Lane for an alternative simple expression for the number of genotypes which can climb to the global optimum: Consider a genotype on a random landscape which is able to descend to 2 less fit variants, each of which on average can descend to a single still less fit variant. That initial genotype is able to descend to 5 less fit variants. But its predecessor in descending from the global optimum had, on average, 4 less fit variants, and the next higher predecessor had 8 less fit variants, and so on until the descent started at the global optimum. This leads to:

$$5/2 \times 2^1 \times s^2 \times \ldots \times 2^M = 5/2 \times 2^{M(M+1)/2}$$

$$\approx 5/2 \times 2^{M^2/2} \tag{12b}$$

where $2^M = D$.

In this notation, the total number of local optima, $(2^D/N + 1)$, is $(2^{2^M}/2^M + 1)$, which is vast compared to $5/2(2^{M^2/2})$ even for modest M.

8. Conflicting constraints cause a "complexity catastrophe": as complexity increase accessible adaptive peaks fall toward the mean fitness

We now investigate the inexorable onset of a novel complexity catastrophe which limits selection. It is the consequence of attempting to optimize in systems with increasingly many conflicting constraints among the components: Accessible optima become even poorer. Fitness peaks dwindle.

In the NK model, K measures epistatic interactions. We have already noted that as K increases, this increases the number of conflicting constraints. Thus, if $K = 0$, each gene can assume its most valuable allele independently of the choice at any other allele. If $K = 1$ and genes I and J mutually influence one another, the optima choice of allele for I in the context of all possible choices of alleles at J, will typically not be identical to the optima allele at J for all possible alleles at I. These conflicting constraints mean that the best mutual choices of alleles tend to be poorer overall. As K increases, the web of constraints becomes enormously complex. When K increases to $N - 1$, we might expect the conflicts among constraints to reach a maximum, and hence to poorer local optima than for smaller values of K relative to N. Indeed this proves to be the case. We examine this "complexity catastrophe" first in the limit of fully random landscapes, when $K = N - 1$, as N grows larger. Below I show that the tendency for local optima to fall in fitness occurs for a range of NK landscapes in which K increases as a constant fraction of N.

To be concrete, and to make the general argument more transparent, I will modify the NK model slightly and assign fitness values at random, not from the *uniform* distribution, but from the extreme of a *U shaped distribution* having only the extreme values in the range 0.0 and 1.0. Also for simplicity I consider the case where each gene has 2 alleles. For each genetic locus, either a fitness contribution of 0.0 or of 1.0 is assigned to each of the 2^N combinations of the 2 alleles of the $K + 1 = N$ loci bearing on the fitness contribution of that locus. As before, the fitness of any genotype is the average of the contributions from all N loci. Thus, for any genotype, its *fitness* is just the fraction of the N loci whose fitness contributions are 1.0. Therefore, the distribution of fitness values among the genotypes is now just the binomial distribution of the sum of N random variables X, where $X = 0.0$ or 1.0. The mean of the distribution is 0.5 and the distribution approaches Gaussian rapidly as N increases.

Note that I am no longer considering a landscape in which fitness is only rank ordered, but one in which the fitness values are real valued and are drawn at random from a defined range, 0.0 to 0.1

A simple way to think about the lengths of adaptive walks is that they continue until the expected number of fitter 1-mutant neighbors drops below 1. Since each genotype has N neighbors, on average, walks stop when the expected fraction of fitter neighbors falls just below $1/N$.

The fitness of a genotype, f, is the fraction of its N loci which make fitness contributions of 1.0. Because the fitness of 1-mutant neighbors are random, the expected proportion of 1-mutant neighbors which have higher fitness than that of a genotype is simply given by the *probability* in the "right tail" of the binomial distribution above its fitness. As the fitness of genotypes, f, increases, say from 0.6 to 0.7, the probability above this increased fitness value decreases. Walks will continue to higher fitness values until the expected fraction of fitter neighbors falls to $1/N$.

We now show that as N increases, the fitness values of attainable local optima decrease toward .5. For N large the central limit theorem shows that the fitness has approximately a normal distribution with mean $1/2$ and variance $1/(4N)$. Using the above arguments, the expected fitness, f, of an attainable local optimum is found from

$$\Pr(x > f) = 1/N \tag{13}$$

where x has a normal distribution with mean .5 and variance $1/(4N)$. If the distance of a local optimum above .5 is $f^* = f - .5$, this gives, using the approximation

$$\int_y^\infty \frac{1}{\sqrt{2\pi}} e^{1\frac{1}{2}u^2} du \\ \sim \frac{e^{-\frac{1}{2}v^2}}{y\sqrt{2\pi}} \tag{14}$$

the following equation for f^*:

$$\frac{.199(N)^{\frac{1}{2}}}{f^*} \cdot e^{-2(f^*)^2 N} = 1. \tag{15}$$

This expression is adequate for reasonably large values of N.

Equation (15) implies that as the number of genes, N, increases, the accessible optima dwindle towards the average unselected fitness in the space of genotypes. The decrease in the fitness of attainable local optima is rapid at first, then more gradual, as N increases. Thus, inexorably in these landscapes, as N increases, adaptive walks terminate on poorer "solutions".

I believe this to be a genuinely fundamental restraint facing adaptive evolution. As systems with many parts increase both the numbers of those parts and the richness of interactions among the parts, it is typical that the conflicting design constraints among the parts increase massively. Those conflicting constraints imply that optimization can only attain ever poorer compromises. No matter how strong selection may be, an adaptive process cannot climb higher peaks than afforded by the fitness landscape itself. That is, this limitation cannot be overcome by stronger selection.

It might be objected that *normalization* is essential to this limitation on selection, and that such normalization is an arbitrary assumption of the model. Yet if we include "costs per part", it is clear that some form of something like normalization is a natural and general consideration. Further it is clear that conflicting constraints are a very general limit in adaptive solution. Each "part" of a complex system costs something. For example, additional genes and proteins require metabolic energy.

As a concrete example, suppose we do not normalize fitness in the NK model and consider the total fitness of the system, ignoring any cost per part. Then as N increases, both the total complexity, N, and the maximum possible fitness, N, increase *without bound*. Nevertheless, increasing total costs will typically bound the overall fitness which can be achieved. In the $K = N - 1$ case, as N increases total fitness increases, albeit ever more slowly. But suppose cost per part is constant, hence total cost arises linearly. At some point, total cost exceeds total fitness. Further increase in complexity, increasing N, is no longer profitable. This shows that there is again a limit on the complexity which can be attained. The "marginal" increase in fitness for the next part must be positive. The complexity catastrophe due to conflicting constraints captured in Eq. (15) is therefore a simple and rather general property of complex systems.

The "Tunable" NK Family of Correlated Landscapes

The NK model was invented not to explore the two extreme landscapes, but to have in hand a model which allowed construction of an ordered family of tunably correlated landscapes.

In order to investigate the statistical properties of landscapes for different values of the fundamental parameters of the NK model, numerical simulations were carried out (Kauffman *et al.* 1988, and Weinberger 1989). The main parameters of the model are N, K, the distribution of K among the N, and A, the number of "alleles" at each site. Other parameters include the underlying distribution from which "fitness" values are assigned to each site for each combination of alleles at the $K + 1$ sites bearing on it. The properties we investigated include:

1) The fitness of local optima.
2) The lengths of adaptive walks to optima.
3) The dwindling fraction of fitter neighbors at each step along an adaptive walk.

4) The mean waiting time to find a fitter variant.
5) The number of local optima.
6) The similarity or "distance" between local optima.
7) The number of adaptive walks from different initial genotypes which climb to each local optimum (hence the "attracting" basin size of each optimum.)
8) The autocorrelation of the fitnesses encountered along a random walk in the landscape as a measure of the correlation structure of the landscape.

To investigate questions 1) – 4), numerical simulations were carried out for different random examples of the NK landscapes for fixed N and K values, initiating adaptive walks from a random initial genotype, then hill climbing via a randomly chosen fitter 1-mutant variant of each successively fitter genotype to the nearest optimum. Walks were carried out on 100 different randomly chosen landscapes for the same values of the model parameters. The number reported are the means of those 100 simulations for each value of the model's parameters.

Perhaps the most surprising feature of the results is that most aspects of these landscapes are so nearly insensitive to any parameters but N and K.

Table 1 shows the average fitness of local optima attained in the case where the K epistatic inputs to each gene were chosen to be its flanking $K/2$ neighbors to either side. To avoid "boundary effects" we considered "circular" genomes. Each gene was limited to $A = 2$ alleles, 0 and 1.

Table 1. Mean fitness of local optima (nearest neighbor interactions).

		N			
	8	16	24	48	96
0	0.65(.08)	0.65(.06)	0.66(.04)	0.66(.03)	0.66(.02)
2	0.70(.07)	0.70(.04)	0.70(.08)	0.70(.02)	0.71(.02)
4	0.70(.06)	0.71(.04)	0.70(.04)	0.70(.03)	0.70(.02)
8	0.66(.06)	0.68(.04)	0.68(.03)	0.69(.02)	0.68(.02)
K 16		0.65(.04)	0.66(.03)	0.66(.02)	0.66(.02)
24			0.63(.03)	0.64(.02)	0.64(.01)
48				0.60(.02)	0.61(.01)
96					0.58(.01)

Table 1 shows first that for $K = 0$ the fitness of optima are independent of N and are about 0.66, or 2/3. This is expected. Fitness values are drawn at random between 0.0 and 1.0 for allele 0 and allele 1. Order statistics shows that the average value of the less fit allele will be 1/3 and the more fit allele will be 2/3. Since each site contributes additively to the overall fitness which is merely the mean fitness per site, the global optimum should be independent of N, and should be 2.3. By a similar argument, for an arbitrary, but fixed number of alleles, A, the average fitness of the fittest of these, if drawn at random from the uniform distribution between 0.0 and 1.0, is $A/(A + 1)$.

Table 1 shows that in the $K = N - 1$ limit the "*complexity catastrophe*" does certainly occur. The fitness of accessible local optima begin high and dwindle toward 0.5 as N and K increase.

In order to investigate whether the same complexity catastrophe would occur, as expected, regardless of whether the fitness contributions per site were drawn from the uniform interval between 0.0 and 1.0 or some other distribution, we investigated a V shaped distribution which favored values nearer 0.0 or 1.0, and an inverted V or "humped" distribution which favored decimals near 0.5. In all cases, as expected, in the uncorrelated landscape limit with $K = N - 1$, the fitness of optima recede toward 0.5 as N and K increase.

Three further features of Table 1 are important. First note that if K is small and fixed ($K = 2$, $K = 4$, or $K = 8$) while N increases, then the fitness of optima do not fall. The complexity catastrophe appears to be averted for small fixed values of K. Second, the fitness of optima for fixed small values of K ($K = 2$ and $K = 4$) is actually higher than for $K = 0$. Thus, low levels of epistatic interaction appear to "buckle" the landscape rather like heaving up mountain ranges, and yield fitter optima than those available in the simplest additive "Fujiyama" landscape. Third, inspection of the table shows that as K increase as a constant proportion of N then the fitness values may transiently increase but ultimately *fall toward 0.5*. This third point demonstrates that if K increases linearly with N the *complexity catastrophe* sets in. Therefore there is some very broad set of landscapes within the NK family subject to the limitation that optima recede to the mean of the space as N increases.

These behaviors of the fitness of optima as a function of N and K suggest that two major regimes within the NK family of landscapes exist: 1) K remains small, of order 1, as N increases. 2) K grows with N, hence is of order N, as N increases. In the latter case the complexity catastrophe arises. In the former case, optima remain high. Presumably these two regimes are different in many basic respects which require investigation.

As walks proceed toward optima the number of fitter 1-mutant neighbors dwindles to 0. The reciprocal of the fraction of fitter 1-mutant neighbors is the expected waiting time to find such a fitter variant. Figure 5a shows the logarithm of the average number of fitter mutants, and Fig. 5b shows logarithm of the waiting time for values of $N = 96$ as K increases for the neighboring K cases. Data for K random are similar. Note that for $K = 0$ the fraction of fitter neighbors dwindles slowly, and as K increases the fraction dwindles ever more rapidly. Further, for $K = 2$ or more, the fraction of fitter neighbors falls off by approximately a constant fraction of each improvement step. That is, the falloff in fraction of fitter neighbors is approximately exponential for $K >= 2$. The slope increases as K increases, toward log of the expected limit of a fully random landscape. That the slope is approximately log linear even for $K = 2$ is rather interesting, for it suggests that this is a quite general feature of rugged landscapes even when those landscapes remain quite highly correlated.

Table 2 shows similar results, but under the condition that the K genes which epistatically affect each locus were chosen entirely at random for each locus. The constraint to circular genomes is removed, and no reciprocity in epistatic influence is assumed. The main feature of Table 2 compared to Table 1 is that it is nearly the same. Therefore, within the NK family of correlated landscapes, the actual fitness of optima are largely insensitive to the distribution of the K among the N.

Table 3 shows the average number of steps via fitter 1-mutant variants for the NK model with K drawn from neighbors as in Table 1. For $K = 0$, walk lengths to optima are about $N/2$. For $K = N - 1$, walk lengths are close to $\log_2 N$. The deviation may reflect sampling effects. Thus walk lengths vary from linear in N to logarithmic in N as K increases. Table 4 shows similar data for the randomly chosen K case of Table 2. Again, Tables 3 and 4 are very similar. Thus whether the K epistatic "inputs" to a gene are its neighbors or random among the N has almost no bearing on the lengths of walks to optima.

We also examined the *ruggedness* of the fitness landscape in the 1-mutant vicinity of local optima. More precisely, walks were carried to local optima, then the fitness of all 1-mutant variants of such optima were assessed. Our expectation is that as K increases the landscapes become

Table 2. Mean fitness of local optima (random interactions).

		N				
		8	16	24	48	96
	2	0.70(.06)	0.71(.04)	0.71(.03)	0.71(.03)	0.71(.02)
	4	0.68(.05)	0.71(.04)	0.71(.04)	0.72(.03)	0.72(.02)
	8	0.66(.06)	0.69(.04)	0.69(.04)	0.70(.02)	0.71(.02)
K	16		0.65(.04)	0.65(.03)	0.67(.03)	0.68(.02)
	24			0.63(.03)	0.65(.02)	0.66(.02)
	48				0.60(.02)	0.62(.02)
	96					0.58(.01)

Table 3. Mean walk lengths to local optima (nearest neighbor interactions).

		N				
		8	16	24	48	96
	0	1.5(1.2)	8.6(1.9)	12.6(2.2)	24.3(3.4)	48.8(4.6)
	2	4.1(1.9)	8.1(3.2)	11.2(3.1)	22.5(4.6)	45.2(6.6)
	4	3.2(1.8)	6.6(2.5)	9.4(2.9)	19.3(3.9)	37.3(6.1)
	8	2.7(1.5)	4.7(2.3)	7.7(3.0)	15.3(4.3)	27.7(5.3)
K	16		3.3(1.7)	4.8(2.1)	9.6(3.0)	19.3(4.2)
	24			3.5(1.4)	7.4(3.0)	5.0(3.9)
	48				3.9(1.9)	8.9(3.0)
	96					5.1(2.4)

Table 4. Mean walk lengths to local optima (random interactions).

		N				
		8	16	24	48	96
	2	4.4(1.8)	8.1(2.8)	12.5(3.8)	26.5(5.1)	46.9(6.1)
	4	3.6(1.8)	7.3(2.9)	10.9(3.3)	22.9(5.6)	44.5(7.9)
	8	2.7(1.5)	5.3(2.5)	8.0(3.2)	17.0(4.3)	34.7(6.5)
K	16		3.3(1.7)	4.8(2.1)	10.1(3.4)	21.6(4.8)
	24			3.5(1.4)	7.4(2.6)	16.0(4.3)
	48				3.9(1.9)	9.3(2.6)
	96					5.1(2.4)

progressively less correlated. Therefore the fitness "drops" away from optima for high values K should tend to be more precipitous than for the gentler landscapes with low values of K. This is clearly seen. More rugged landscapes generally also are more rugged in the vicinity of local optima. This implies that the selection gradients back to optima are steeper in rugged than in gentle landscapes. This will clearly effect how populations adapt on such landscapes, as we discuss below. Similar results are found for K random landscapes as K increases.

Questions 5)–7) above, concerning the number of local optima, the distance between local optima, and the sizes of basins of genotypes climbing to each local optimum, are *non-local* features of NK landscapes. To investigate these properties we randomly chose a specific NK landscape and carried out many adaptive walks to local optima from randomly chosen initial genotypes in the space. To establish the number of local optima, searches were carried out until no further local optima were uncovered, or until 10,000 optima were discovered. Any such algorithm has the difficulty that optima with very small basins of attraction may be missed. The sampling processes are therefore biased by the distribution of sizes of basins. Therefore, current numerical values are estimates. Table 5 shows the results for adjacent and random choices of K and different values of N and K. Values given are the means of three landscapes.

Table 5. Number of optima.

		N	
		8	16
	2	5	26
	4	15	184
K	7	34	—
	8	—	1109
	15	—	4370

A Massif Central in $K = 2$ Landscapes

Among the most surprising features of the NK family of landscapes is the fact that for small values of K and two alleles, the local optima are not distributed randomly in genotype space, but instead are *near* one another. Thus there is a *global structure to the fitness landscape*. More precisely, *the highest optima are nearest to one another*. The natural measure of the distance between two genotypes with only two alleles per locus, 1 or 0, is the *Hamming distance*. The Hamming distance is simply the number of the N positions at which the alleles differ. Thus (00000) and (10000) have a Hamming distance of 1. If local optima were distributed randomly in genotype space, the average Hamming distance between two local optima would be $N/2$. Furthermore, the Hamming distance from the highest local optima, i.e., the fittest found, to the second highest local optimum would on average be $N/2$. This is clearly not the case for small values of K, whether the K epistatic inputs per site are adjacent to that site, or even randomly distributed among the N. For K small, e.g., $K = 2$, the highest optima are nearest one another. Further, optima at successively greater Hamming distance from the highest optimum are successively less fit. Thus as stated, there is a global order to the landscape. Like the Alps, it possesses a kind of "Massif Central" or high region of genotype space where all the good optima are located. As K increases, this correlation falls away, more rapidly for K random than K adjacent. The "Massif Central" is the analogue of a dynamical "frozen component" in parallel processing Boolean networks, where a frozen component itself corresponds to a set of sites in states which is mutually consistent, hence not *frustrated*.

The NK family of landscapes offers further surprises: The distribution of basin sizes climbing to specific optima can be very non-uniform. Some basins are enormous. Further, for K small, there is a tendency for the *highest optima to have the biggest basins*. Thus, simultaneously, the global order of the "Massif Central" in this landscapes is expressed by the fact that the highest optima are both nearest one another, and have the largest drainage basins. In turn this implies that one high local optimum has *information* about where other good local optima are. And, further, the region *between* two high local optima is a good area to search for still higher local optima. This mutual information carries marked implications about the usefulness of *genetic recombination* as a search strategy in such rugged but correlated fitness landscapes.

Other Combinatorial Optimization Problems and Their Landscapes

The general purpose of this chapter has been to discuss the *structure of rugged fitness landscapes*. I have done so in terms of the NK model. I bring the general discussion of the structure of rugged landscapes to a close by mentioning a few other examples of complex combinatorial optimization problems. Each has a complex fitness or cost landscape.

I have already briefly mentioned spin glasses. These magnetic materials have afforded a rich source of models of complex potential surfaces, (Edwards and Anderson 1975, Sherrington and Kirkpatrick 1975, Anderson 1985, Binder and Young 1986). Analysis of these systems have explored topics such as the number of global minima, the spin overlaps between minima, the height of potential barriers between minima, and slow relaxation times at finite temperatures as a spin glass explores its potential surface.

The close relation between the NK model and spin glasses again warrants comment. As noted, the NK model is a version of a spin glass in which the fitness contribution of each site is written as a sum of terms depending upon the "allele" at that site, plus terms for pairwise interactions of that allele with each of the K others impinging upon it, plus more terms for all the triadic combinations among the $K + 1$ sites, plus terms for all combinations up to the "K-adic". This is closely related to a spin-glass model investigated by Gross and Mezard (1985) which examines a Hamiltonian energy surface given by the sum of all possible K-adic interactions among the spins. The Gross-Mezard model (1985) appears to be almost identical to a model introduced by Amitrano *et al.* (1989) to study molecular evolution on rugged landscapes. The limit of the Gross-Mezard spin glass when K approaches N is Derrida's (1981) random energy spin glass model. In fact the random energy model is identical to the NK model in the $K = N - 1$ limit. Thus results we have obtained for this limiting case apply to the corresponding model in statistical physics as well.

The NK family of correlated landscapes is but one family of landscapes. It is an entirely open question whether there may be a few fundamental families of correlated landscapes, or an extremely large number of such families. This is a critical question for further investigation. The grounds to hope that there may be a few fundamental families of correlated landscapes is that in many areas of statistics a few fundamental distributions have proved important. Should we eventually discover that rather few families of correlated landscapes exist, then it might prove possible to measure a few parameters of a given rugged correlated landscape and discern both the family to which it belongs, and also how adaptive strategies based on mutation and selection, or otherwise, might best optimize on that particular landscape. But whatever the answer may prove to be, the utter simplicity of the NK family lends credence to my impression that it is likely to be a fundamentally important family of correlated landscapes.

There are, in fact, clues that many complex combinatorial optimization problems can be mapped by a few parameters onto the NK family of rugged landscapes. Weinberger (private communication) has recently applied an autocorrelation measure of fitness landscapes to an apparently very different problem, that of RNA folding stability for model RNA sequences. Weinberger found that the autocorrelation structure of the model RNA stability landscape corresponded, via the NK model, to an effective K of 8 for RNA sequences length $N = 70$. Utilizing these parameters, via the NK model, he was able to predict the actual observed matter of local optima in model RNA sequence space for folding stability. This suggests that the autocorrelation function, or other measures, may allow many apparently dissimilar problems to be mapped onto the NK family of landscapes. If so, its statistical properties can be used to predict features of other complex cost surfaces.

Summary

This chapter has introduced the concept of rugged fitness landscapes. Such landscapes undoubtedly underlie adaptive evolution at the molecular and morphological level. In order to study the structure of such landscapes I have carefully eschewed discussion of the actual fitness of whole organisms in their environment. I have instead defined fitness narrowly, as, for example, the capacity of peptides to bind to a specific antibody molecule, or of proteins to catalyze a specific reaction under standard conditions.

The major discussion of the chapter has explored the expected structure of discrete fitness landscapes corresponding to sequence spaces, such as protein space. Here, adaptive evolution can be considered as an adaptive walk from proteins of low fitness for a specific function towards proteins with high fitness for that function. In the simplest case, adaptive walks proceed via fitter 1-mutant variants to local optima in the space of possibilities. Our analysis has focused upon the statistical features of such walks as a function of the ruggedness of fitness landscapes. Important aspects of adaptive walks include the number of local optima, the number of adaptive steps on an adaptive walk to a local optimum, the average number of mutants "tried" on such a walk, the average number of mutants "accepted" on such a walk, the rate at which fitter 1-mutant neighbors dwindle to 0 along an adaptive walk as local optima are attained, the average number of local optima which are accessible by alternative walks from a starting point, the number of points which can climb to the global optimum, the relative locations of local optima with respect to one another and the global optimum, and the basin sizes "draining" to local optima as a function of the height of that optimum.

We have found that all of these properties differ as a function of how rugged and multipeaked the fitness landscape is. In particular, I introduced the NK family of fitness landscapes to explore these issues. The NK family of landscapes is among the first studied which explicitly invites us to explore the statistical structure of fitness landscapes. In this model, tuning the epistatic coupling parameter K relative to N, increases the ruggedness of the landscape in a controlled manner from single peaked and smooth for $K = 0$ to fully random for $K = N - 1$.

I have developed and presented the NK model in detail for the following purposes.

There can be no doubt whatsoever that the real adaptive evolution of proteins for specific catalytic tasks, ligand binding tasks, and otherwise, confront fitness landscapes with some statistical characteristics. If the NK model were to serve no other purpose than to tune our intuitions about what such landscapes might look like, that alone would warrant our attention.

In the interpretation of the NK model as proteins, the NK family of landscapes may or may not prove to be useful in actually predicting the structure of real fitness with respect to protein evolution. In fact, the model has been applied to predict features of the immune response with some success (Kauffman 1991).

The NK family already has led us to a clear and largely unrecognized feature of many complex combinatorial optimization processes. As the number of conflicting constraints increases, the corresponding landscapes tend not only to become more multipeaked and rugged, but for an unknown and very large family of landscapes the actual peaks recede toward the mean fitness in the space of possibilities. This is clearly important. Organisms are complex. Do they avoid this limitation? If so, how? If not, what are the implications?

The NK model suggests at least one means to mitigate the complexity crisis. If the number of epistatic interactions, K, remains small while N increases, landscapes retain high accessible local optima. This is a first hint of something like a *construction requirement* to make complex systems with many interacting parts which remain perfectible by mutation and selection. Each part should directly impinge on rather few other parts. A requirement that "K" be low finds resonance in the suggestions of other workers made on rather different grounds. For example, computer scientists and economist H. Simon (1962) has argued for many years that solvable problems are "near-decomposble". By this Simon has meant that tasks which were composed of subtasks which might be solved in relative isolation, then recombined, were far easier to solve than those whose parts were richly interconnected. It is a familiar experience with machines that we typically design systems in which each part interacts directly with rather few other parts.

However, the story is more complex. Optimization of landscape structure to optimize adaptation is subtle. The complexity catastrophe we have identified is averted in the NK model for those landscapes which are sufficiently smooth to retain high optima as N increases. But we have ignored discussion of the actual *flow of an adapting population* under the simultaneous drives of mutation and selection. We might wonder whether averting the present complexity catastrophe has a price. In fact, there is such a price. As landscapes become *smoother*, the fitness gradients toward adaptive peaks becomes shallower and therefore selection is less able to hold adapting populations at the gentler peaks. In the fact of a fixed mutation rate, as systems under selection increase in complexity, a *complexity or error threshold* is reached. Beyond it, populations accumulate mutations and flow away from optima down into the lowlands of the fitness landscape. Solving the conflicting constraints complexity catastrophe is accomplished at the price of making selection into a weaker Demon!

Scylla and Charybdis were the twin perils which flanked Ulysses. Similar perils flank selective evolution. Given that the structure of fitness landscapes governs the adaptive process, we must consider the possibility that natural selection itself "tunes" landscapes structure to optimize adaptive evolution. But the task is not simple. There are conflicting reasons why it might be advantageous for fitness landscapes to be smooth or to be rugged. Optimizing ruggedness requires compromise. Tuning landscapes smoother leaves the peaks higher, but lessens the capacity of selection to pull populations up the smoother hills. Conversely, it can be advantageous for landscapes to be more rugged. When they are, the *sides* of fitness peaks are steeper, allowing more rapid increase in fitness for a modest change in genotype. This can be advantageous in the face of a higher mutation rate, or in the presence of fitness landscapes which *deform* due to changes in the abiotic environment, or to coevolutionary processes. But as landscapes become more rugged, adaptive processes tend to become trapped into smaller regions of the space, thereby hindering the capacity to evolve, and the increase in conflicting constraints tends to lower the adaptive peaks.

References

Alberts, A., Bray, D., Lewis J., Raff, M., Roberts, K. & Watson, J. D. (1983), *Molecular Biology of the Cell* (Garland, New York, [3, 5]).

Amitrano, C., Peliti, L., and Saber, M. (1988), "Population dynamics in a spin-glass model of chemical evolution", submitted to *J. Mol. Evol.*, Dec. 1988.

Anderson, P. W. (1985), "Spin glass Hamiltonians: A bridge between biology, statistical mechanics, and computer science", in *Emerging Synthesis in Science-Proceedings of the Founding Workshops of the Sante Fe Institute*, ed. David Pines, The Santa Fe Institute, Santa Fe, New Mexico. [3, 8].

Borstnik, B., Pumpernik, D. and Hofacker, G. L. (1987), "Point mutations as an optima search process in biological evolution", *J. Theor. Biol.* **125**, 249.

Crow, J. R., Kimura, M. (1970), *An Introduction to Population Genetics Theory* (Harper and Row, New York, [3]).

Crow, J. R., Kimura, M. (1965), "Evolution in sexual and asexual populations", *Am. Nat.* **99**, 439.

Derrida, B. (1981), "Random energy model: An exactly solvable model of disordered systems", *Phys. Rev.* **B24**, 2613.

Eigen, M. (1985), "Macromolecular evolution: Dynamical ordering in sequence space", in *Emerging Synthesis in Science — Proceedings of the Founding Workshops of the Santa Fe Institute*, ed. David Pines, The Santa Fe Institute, Santa Fe, New Mexico, pp. 25–69.

Eigen, M. (1987), "New concepts for dealing with the evolution of nuclear acids", in *Symposia on Quantitative Biology*, Vol. LII, Cold Spring Harbor Laboratory, pp. 307–320.

Eigen, M. and McCaskill, J. (1989), "The molecular quasi-species", *Adv. Chem. Phys.*, in press.

Ewens, W. (1979), *Mathematical Population Genetics* (Springer, New York).

Feller, W. (1971), *Introduction to Probability Theory and Its Applications*, Vol. II, 2nd ed. (Wiley, New York).

Fontana, W., Schuster, P. (1987), "A computer model of evolutionary optimization", *Biophys. Chem.* **26**, 123.

Franklin, I., Lewontin, R. C. (1970), "Is the gene the unit of selection?" *Genetics* **65**, 707. [2]

Gillespie, J. H. (1983), "A simple stochastic gene substitution model", *Theor. Pop. Biol.* **23**, 202.

Gillespie, J. H. (1984), "Molecular evolution over the mutational landscape", *Evolution* **38**, 1116.

Jacob, F. (1983), "Molecular tinkering in evolution", in *Evolution from Molecules to Men*, ed. D. S. Bendall, (Cambridge University Press, Cambridge), pp. 131–144. [3]

Kauffman, S. A., Levin, S. (1987), "Towards a general theory of adaptative walks on rugged landscapes", *J. Theor. Biol.* **128**, 11. [3, 4]

Kauffman, S. A. (1992), *Origins of Order* (Oxford Univ. Press, New York), to appear.

Kauffman, S. A. and Stein, D. (1989), "Application of the NK model of rugged landscapes to protein evolution and protein folding", Abstract AAAS Meeting on Protein Folding, June 1989.

Kauffman, S. A. and Weinberger, E. D. (1989), "The N-K model of rugged fitness landscapes and its application to maturation of the immune response", submitted to *J. Theor. Biol.*

Kauffman, S. A., Weinberger, E. D., Perelson, A. S. (1988), "Maturation of the immune response via adaptive walks on affinity landscapes", in *Theoretical Immunology*, Part I, SFI Studies in the Sciences of Complexity, ed. A. S. Perelson, (Addison-Wesley), pp. 349–382. [3, 4]

Levin, S. A. (1976), "On the evolution of ecological parameters", in *Ecological Genetics: The Interface*, ed. P. F. Brussard (Springer, New York), pp. 3–26.

Levine, M. and Harding, K. (1987), "Spatial regulation of homeo box gene expression in *Drosophila*", in *Oxford Surveys of Eukaryotic Genes*, ed. N. MacLean, (Oxford Univ. Press, Oxford).

Lewontin, R. C. (1974), *The Genetic Basis of Evolutionary Change* (Columbia Univ. Press, New York). [2]

Macken, C. A. and Perelson, A. S. (1989), "Protein evolution on rugged landscapes", *Proc. Natl. Acad. Sci. USA,* submitted.

Maynard Smith, J., (1970), "Natural selection and the concept of a protein space", *Nature* **225**, 563.

Ninio, J. (1979), "*Approaches moleculaires de l'evolution*", *Collection de Biologie Evolutive* **5**, 93.

Provine, W. B. (1986), *Sewall Wright and Evolutionary Biology* (The University of Chicago Press, Chicago).

Schuster, F. (1986), "The physical basis of molecular evolution", *Chemica Scripta* **26B**, 27.

Schuster, P. (1987), "Structure and dynamics of replication-mutation systems", *Physica Scripta* **35**, 402.

Simon, H. (1962), "The architecture of complexity", *Proc. Am. Phil. Soc.* **106** 467. [5]

Weinberger, E. D. (1988), "A more rigorous derivation of some properties of uncorrelated fitness landscapes", *J. Theor. Biol.* **134**, 125.

Weinberger, E. D. (1990), "Correlated and uncorrelated fitness landscapes and how it tells the differences", *Biol. Cybernet.*, in press.

Wright, S. (1932), "The roles of mutation, inbreeding, crossbreeding and selection", in *Evolution — Proceedings of the Sixth International Congress on Genetics*, pp. 356–366.

Wright, S. (1931), "Evolution in Mendelian populations", *Genetics* **16**, 97.

Self-Organization in Prebiological Systems: A Model for the Origin of Genetic Information

D. S. Rokhsar

Department of Physics, University of California,
Berkeley, CA 94720, USA

This paper addresses the problem of biological information by introducing a measure of its value inspired by spin glasses. A model for the origin and selection of biological information is then discussed, first from the point of view of a general theory, and later with respect to a collection of computer simulations of a specific model. This model demonstrates some interesting properties of autocatalytic systems, establishing its plausibility as a model for prebiological systems.

1. Biology and Information

1.1. Physics and Biology

The elementary constituents of life are ordinary atoms like those that make up rocks and other non-living objects, yet living systems possess complex properties and behavioral characteristics that are not even hinted at in inanimate matter. Naively, this complexity is surprising: The fundamental laws that govern nature on microscopic scales are simple, exhibiting symmetries which are not evident in most macroscopic systems. To the modern physicist, however, the possibility that microscopic degrees of freedom will order themselves to generate new properties at the macroscopic level is a familiar one. Characteristics which appear in a system but are not present in its components are known as emergent properties,[1] and the study of the murky regime during the "crossover" from the inanimate to the living state must be the study of the emergence of these properties.

Our consideration of these and other questions will emphasize the transfer and utilization of information as we discuss the meaning and value of information in a biophysical context, and we apply these concepts to a simple model for the origin of biological information. Our model consists of a collection of self-replicating polymers, which can exhibit the prebiological properties we seek. This model has been simulated on a computer, and numerical experiments have been performed to determine the usefulness of the model. Throughout our discussion we will use the concepts of condensed matter physics to express our (hopefully not too naive) thoughts on these matters.

Any simplified scenario for the "origin of life" tacitly uses the concept of the *universality* class,[3] a collection of systems that exhibit the same essential behavior (for example, common critical exponents at a phase transition) despite differences in the details of their operation (i.e., microscopic makeup). Again, this is not to say that the details of life on earth are unimportant, but merely reflects the physicists' prejudice that parameters should not have to be finely tuned and should be relatively insensitive to biochemical details. For example, the use of polynucleotides as information carriers appears to be universal, but the fundamental property of such molecules is their ability to store information and replicate, not the particular bonding mechanisms that are used in these processes. (Alexander Rich[2] has even suggested that synthetic information bearing polymers will eventually be developed which will lead us to a better understanding of our particular version of life.) By proceeding in this manner, we can hope to make a simplified model that will exhibit some of the fundamental properties of living matter, i.e., a model which is in the same universality class as the vastly more complicated process which actually led to the appearance of life on earth.

A great deal of research has been done on the chemical details of the origin and evolution of life, and most of this work can be placed into two categories — the ability of the non-living environment to produce the small biomolecules needed as precursors to larger biological structures, and the evolution of microorganisms to complex organisms like ourselves. The pioneering work of Oparin, Urey and Miller, and others (for a review of this research, see Fox[4]) has demonstrated that the primordial oceans could have produced a supply of amino acids and nucleic acids, driven by solar ultraviolet radiation and various terrestrial energy sources. Darwin and Wallace's principle of natural selection, interpreted via modern genetics, can lead from the first microorganisms to more sophisticated forms of life. Our concern is with the time period between these two stages of evolution: The emergence of living systems from a non-living origin. In order to accomplish this goal, we must first look for the emergent properties that characterize life.

1.2. What is Life?

In his essay "What is Life?" Schrödinger described life by appealing to its nonequilibrium nature. "Living matter evades the decay to equilibrium ... by continually drawing from its environment negative entropy". In order to preserve the highly ordered structures that characterize life, organisms must be constantly dissipating entropy into their environment, thereby utilizing "negative entropy" from the surroundings. A primary feature of life is its dissipative character, since all life has a metabolism that enables the organism to process the energy and negentropy of its surroundings to preserve its structure. A "metabolism" is the first of our requirements for life.

Yet another important feature of life is its long and short term stability. In the short term, a viable organism must be able to alter its behavior to respond to fluctuations in the environment, preserving itself as a region of low entropy. Present day life is at such an advanced level that most organisms can survive under a large range of conditions, having evolved homeostatic mechanisms through millions of years of evolution. The first "living" systems were surely not as flexible, lacking these complex regulatory mechanisms. The metabolisms of these protoorganisms must have been correspondingly simple; a reasonably stable macromolecule would meet our criterion of short term stability with respect to small scale environmental changes. To give long term stability to a particular "species," or progression of similar organisms, we require a metabolism that will reproduce itself with some reasonable accuracy. The organisms we have been exposed to therefore have two adaptation time scales: A short term metabolism which damps our fluctuations and a longer term (i.e., many lifetimes) adaptation during which the metabolism of the species can change. The two adaptation modes are coupled by the principle of natural selection: Those that adjust best in the short term will be more likely to transmit their particular genotype to future generations.

Life is then a progression of metastable creatures, each able to react efficiently to fluctuations, which can also adapt to long term trends in the environment. Is reproduction a necessary condition for this long term stability? With probability unity any individual organism will eventually encounter an environmental fluctuation it cannot survive, so that immortality of the information content of the organism requires multiplication of copies. (This is identical to the argument that no long-range order can exist in a one-dimensional chain.) Thus long term stability requires reproduction of organisms.

The use of the terms "transmission" and "propagation with reference to reproduction is suggestive of the transfer of information — a metabolism carries with it information concerning its structure. There are a variety of ways to transmit a metabolic machinery, one way being the self-replication of the various parts of the organism. Variation occurs when replication is imperfect; those parts which function better give their hosts an advantage, so that more efficient metabolisms are reproduced. If a structure were efficient but slowly replicating, it might die out, leaving less efficient mechanisms to survive (at least in primitive systems). Such a structure might however not be uncommon, since the criteria for an efficient metabolism could be considerably different from the requirements for accurate and rapid reproduction. Early prebiological systems may well have used this system, since it is considerably simpler than the reproduction mechanism used in present day life, which requires separate structures to translate and store the information used to describe the metabolism.

With separate information storage and utilization, only the information carrrier need be replicated; the information and a structure to decode it are all that are required to define a particular metabolism. The constraints of the self-replication model of the previous paragraph do not appear here, since the metabolism is free to evolve independently of the constraints relating function and replication. Once a reasonably efficient translating apparatus is available, the system can quickly evolve a more efficient metabolism.

An often drawn analogy can be made between living organisms and computer systems concerning the utilization and transport of information. The translation of metabolic information can be likened to the compilation and running of a computer program. The program itself is highly portable, and needs only another computer to produce results. As any computer programmer knows, an error (mutation) can often lead to a program that runs, but which does not perform its task properly. The analogue of a self-replicating system without separate information carriers is an analog circuit, combining instructions and processing in one unit. Such an arrangement lacks the flexibility of a digital computer, since a change of instructions changes the functional form of the circuit, constraining the results.

A model for the "origin of life" must therefore allow for the transition from a disorganized original state to a stable, dissipative, self-reproducing system. For our original state, we assume that the materials for life are available in some activated form, ready for assembly into the metabolic structures of the organism. Note that by including self-replication and stability (for both the individual and species) we exclude most currently understood dissipative structures, owing to their lack of stability.[1] Life can be regarded as succession of organisms that maintain themselves over a long time span. It is this succession of organisms, and the transfer of information between them, that leads to the importance of information carriers in biology. To understand the origin of life in this sense, we must consider mechanisms for information storage and their possible origins.

1.3. Information: Content and Value

In a broad sense, the information *content* of a message can be defined as the additional knowledge that can be gained by receiving it. We define a message in its most general form; a message is any signal, whether it be visual (a picture), aural (a sequence of sound pulses), etc. As the *additional* knowledge acquired, information content depends on the knowledge present before reception of the message. For example, if we know in advance that the message will consist of a string of ten A's and B's, there are 2^{10} equally probable messages. If we are told in advance, that nine of the ten letters will be A's, however, there are only ten possible messages. Clearly, the receipt of a message from the first set of sequences conveys more information than a message from the second set. For equally probable messages, the information content is the logarithm of the total number of possible messages:

$$I = -\log_2 p = \log_2 W , \tag{1}$$

where W is the number of possible messages, and p is the probability of a given message.[6] Defined in this manner, information is additive; two consecutive messages have the same information content as one message which is twice as long because probabilities are multiplicative.

In real messages, the elements of the message are rarely equally probable. In the English language, for example, letters like Z and X are rarely used, and other correlations are produced by the limitations for pronunciation (an M very rarely follows a P, for example). Such constraints always decrease the information content of a string[7] because they effectively limit the number of likely messages. In this case, the information contained in a message with different *a priori* probabilities is

$$I = -N \sum_i p_i \log_2 p_i , \qquad (2)$$

where N is the length of the message and p_i is the probability of the ith symbol. For the English language, correlations and varying letter frequencies lower the information content to less than two bits per letter. This decrease in information is closely related to the added structure and organization that these constraints give to English messages. Most generally, any order that is given to a collection of messages will result in a less efficient transfer of information. It is often said that information is the opposite of entropy; physically, the message is the position of the system in phase space, chosen from the *a priori* region of phase space determined by such conditions as constant energy, etc.

The information content, I, represents the maximum amount of information that can be extracted from a message. What we will call the *meaning* of information is quite different, but is bounded by the content of the message. The meaning of a message can only be defined when a code is associated with a set of messages, so that each message can be translated into a meaningful statement. For example, a degenerate code (i.e., a many to one mapping from the message to its meaning) causes messages to contain much less meaningful information than is theoretically possible. Information content is defined in terms of the likelihood of a given message, while we defined meaning in terms of the probability of a given meaning. Meaning is therefore not a property of the message alone, but also of the structures that interpret the message. The information content in a book is huge if we include all the degrees of freedom present, such as the page numbers on which letters occur, the precise typeset of each letter, etc., but none of these influence the meaning of the text.

A third (and much less well defined) quality associated with a message is the *value* of its information, an entirely different concept from those of content and meaning.[8] For value to be defined, there must be a meaning assigned to each message. Volkenstein defines value of information only for nonequilibrium systems, as the effect of the reaction of a system to the receipt of the message, and uses the particularly illustrative example of a traffic signal. A traffic light yields one bit of useful (meaningful) information: Red or green, stop or go. The waiting and watching drivers form a system which is very unstable with respect to this message, and the receipt of a signal can trigger a rather large reaction among the drivers and automobiles. Whereas meaning and content are naturally measured in bits, there can be no standard unit of value, since it depends upon the particular "trigger system" being considered. Meaning and content are properties of the information storage and retrieval systems, while value can only be defined with reference to a system that will react to the information. Volkenstein has pointed out that this definition of value can even be applied outside of the sciences and engineering: The value of a work of art can be defined as the impact that it makes on an observer, whose emotional and physical responses are triggered by receiving the information. Although value is the least precise quantity associated with information, it is perhaps the most useful concept in the study of the origins of life, because it focuses on nonequilibrium conditions. The concept of value is closely related to the selective value of an information storage system, about which more will be said below.

1.4. Biochemical Information Storage

Information content and meaning can be carried by any spatiotemporal array of signals. For information to persist throughout the lifetime of an organism, however, it should be spatially coded, since temporal messages alone cannot easily be used to store information. We would expect biological information to be stored in a class of molecules which allows both a great variety of combinations (to maximize the efficiency of information storage by increasing the content) and a relatively simple organization, to facilitate the decoding and utilization of the messages. Both of these requirements are met to some degree by many polymer systems. Modern organisms have selected two types of polymers for use as carriers of information: Polynucleotides and polypeptides. These two different molecules are dual representations of the same message. Polynucleotides like RNA and DNA are used primarily for information storage; they have been chosen primarily for their ease of decoding and the possession of certain copying properties. Polypeptides, as will be described below, are the realization of the "meaning" of the information stored in the polynucleotides. Since monomers related to both of these classes of polymers would have been present on the primordial earth, it is likely that the first stages of life were intricately bound up in the development of *value* for these informational strings.

Nucleic acids are composed of various sequences of four bases: Adenine, guanine (the purines), cytosine and thymine (pyrimidines); in RNA the place of thymine is taken by another pyrimidine, uridine. Using the formalism of 1.3, we can easily calculate the maximal information content of such a string. Here $k = 4$, so the information per monomer for equally probable bases is $\log_2 4 = 2$ bits. A special property of nucleic acids is their ability to recognize one another, forming weak (hydrogen) bonds that can join two polymer chains, resulting in a relatively stable structure. These properties of recognition and stability make polynucleotides especially suitable for information storage, and it was the property of complementarity, together with crystallographic data on the structure of DNA, which led Watson and Crick to propose these molecules as the carriers of genetic information.

Owing to the complimentarity between the pairs (A-T) and (C-G), the percentages of C and G are equal, as are the percentages of A and T, and a collection of polynucleotides can be characterized by its [G + C] content, which can differ greatly from organism to organism. These different *a priori* probabilities will alter the information carrying capacity of the strings, yielding a content of

$$I([G + C]) = -[G + C]\log[G + C] - [A + T]\log[A + T] + 1 \qquad (3)$$

where $[x + y]$ is the percent composition of the specified base pair. Less evolved organisms have a larger fraction of [G + C], presumably because the G-C bond is nearly twice as strong than the A-T bond, leading to a more stabilized system. [G + C] concentrations range from $\sim 40\%$ for humans to $\sim 70\%$ for certain bacteria.[9] Higher organisms, needing more information content to define their structure, have evolved more efficient storage procedures by using more information per base. Less developed organisms could not afford the extravagance of an intrinsically less stable carrier, and consequently have higher [G + C].

1.5. Structure, Function and Value

Functionally, polynucleotides alone provide a limited amount of value to an organism, since they cannot form the wide range of structures available to polypeptides. The only area in which modern polynucleotides perform functional roles is in their own replication and decoding. To do this, nucleotides use self-complementary sequences to form three-dimensional conformations, much the same way that a protein uses disulfide bonds to fold back upon itself. The rigid phosphate-sugar backbone of a double stranded nucleotide limits the degree of bending and generally restricts the useful structures which can form. The restricted number of intricate conformations makes polynucleotides ideal for self-replication and information storage and access. Since double stranded RNA remains a one-dimensional system, it can use a relatively simple decoding mechanism. Recent (and some older) discoveries in the biochemistry of RNA have revealed a number of instances of functional roles for smallish RNA molecules. An additional relevant fact is that nearly 90% of the RNA in the cell is tied up in the ribosome in a form whose function is not yet clear. Hence, even in the modern organism RNA tertiary structure has not entirely lost functional ability.

Proteins, on the other hand, have extremely complex conformations, the source of their great diversity and usefulness as catalysts. For information storage and retrieval, however, the fact that some amino acids are necessarily buried inside the protein makes it very difficult to imagine a protein either replicating itself or allowing itself to be read the way that tRNA and ribosomes read nucleotides.

Ignoring these problems with protein information storage for the moment, we recall that the nuclei acid sequences are given meaning by the almost universal genetic code, which associates triplets of bases with amino acids. There are twenty amino acids, so a random sequences of amino acids will have information content $\log_2 20 = 4.32$ per residue. Since information is additive, we see that at least a triplet code is required, since a doublet nucleotide would only provide $2 \log_2 4 = 4$ bits. The meaning of any three consecutive bases or codon is defined by the amino acid for which it codes. Beyond the code that maps triplets into amino acids, there is a more fundamental physical "code" which maps the primary structure of protein (its amino acid sequence) to its tertiary structure (three-dimensional conformation). This code is highly degenerate and very complex, since different proteins can have similar functions, meaning essentially identical active sites.

Volkenstein[8] has partially categorized the value of this code by considering the effect of a codon mutation on the protein being constructed. Amino acids can be classified in several broad groups; hydrophilic and hydrophobic, etc., and the replacement of one member of a group with another leaves the conformation of the protein relatively unchanged. By computing the probability that a codon change will result in a hydrophilic group being replaced by a hydrophobic group and vice versa, Volkenstein has crudely estimated the value of codons, i.e., the reaction of the trigger system (here the protein assembly mechanism) to the information carriers.

As mentioned above, it is possible that early polynucleotides performed dual roles, as both catalysts like modern tRNA, and information carrier, like mRNA or DNA. It is only through such a feedback mechanism, with the structural and informational properties residing in the same molecule, that polynucleotides could have acquired value, because a protein-nucleic acid loop would depend greatly on the code relating the two. Since presumably this code did not develop immediately, we are led to nucleic acids as performing some rudimentary catalytic roles if they are to acquire value. The intricate three-dimensional folding of proteins restricts the extent to which they could serve such a dual purpose.

Thus, for the sake of definiteness and on plausibility grounds we take as an ansatz that RNA alone was the primitive prebiotic system. The computer model to be introduced, however, is probably independent of this particular ansatz, although its basis is clearer in this particular chemical system.

2. A Model for Self-Replicating Polymer Systems

2.1. Self-Instruction and Autocatalysis

The simplified (simplistic?) discussion of biology and information given above suggests that we concern ourselves with simple models of prebiotic systems which deal with the properties of replicating information strings. We will concentrate on the replication of polynucleotide-like molecules, owing to their property of complementarity, and the final section of this section will briefly discuss the relationship between polynucleotide and polypeptide strings. Self-replication requires a molecule to stimulate the production of a copy of itself by recognizing its components and catalyzing the reaction that combines them to form a replica. The model we will discuss is similar to one proposed by Blum,[10] in which a pool of monomers and small polymers is alternately heated and cooled, perhaps by cyclic temperature changes caused by the (then much faster) rotation of the earth. These would have been greater than present day temperature variations because of the composition of the early atmosphere. During the cool phase, molecules would tend to minimize their free energy by forming conformations stabilized by hydrogen bonds (later, we will refer to these weak, complementary bonds as "single" bonds, denoted A-B) with nearly complementary strings. Presumably the *error rate* or probability of a non-complementary hydrogen bond would have been rather large at this stage, without specialized enzymes available to insure the modern accuracy[11] of one part in 10^5. We can expect the uncatalyzed reaction to proceed with an error rate of several percent, since non-complementary pairs can form and would add some stability to the double or multistranded molecule. When the temperature rose in the second half of the cycle, the polymers would dissociate to form single strands once again. If the environment maintains an incoming flux of energy rich monomers, temperature cycling would allow polymers to catalyze the formation of replicas (or complements), since the cool phase would allow the activated monomers to bind selectively to these polymers and then to join covalently (symbolically, a "double" bond A = B) to form a complete complementary strand. The cycle and its important features are diagrammed in Fig. 1.

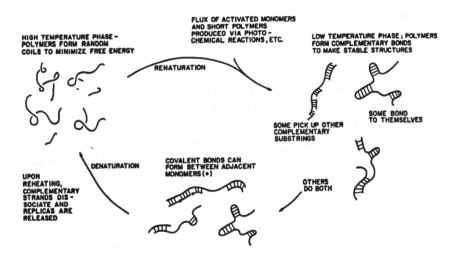

Fig. 1. Temperature cycling for building up longer polymer chains (after Blum).

This cycle resembles repeated renaturation and denaturation of poly-nucleotides. The renaturation step generally proceeds very slowly unless denaturation has been incomplete,[12] so that at least part of the complementary strands remain bound together. This is not the situation that we are interested in, however, since this type of renaturation will only re-bind the complementary copies from the previous generation rather than producing a new polymer from substrings present in the primordial soup. Although the renaturation step we require will be slow, some complementary binding will occur, and replicas will be made. Geometric recognition will be enough to insure at least a similarity between the template and complement strands. One could also conceive of an alternate set of bases that recognized themselves rather than complements, but the cycle would be essentially unchanged; identical copies rather than complementary copies would be formed. On practical grounds such a system is unlikely because there is sure to be thermodynamic bias in favor of either G or C, and complementarity is necessary to ensure degeneracy of the two — i.e., to make sure that our message contains *some* information. This is a primitive version of the "spin glass" argument of the next sections.

In this scheme strands that have fewer self-complementary sections will replicate more rapidly, since they would be less likely to form stable configurations by folding back on themselves, as is the case with tRNA. In particular, palindromic RNA:

...ABCDD'C'B'A'...	palindromic RNA
...A'B'C'D'DCBA...	and its complement

would not replicate well, and would soon become overwhelmed by non-palindromic sequences. It is interesting to note that eucaryotic cells (higher organisms) possess many such sequences, while the simple procaryotic cells do not. This indicates that palindromic sequences only become available after the development of rudimentary polynucleotide replication and the first organisms, so that any selective advantage which palindromes possess would have been able to develop only in the less demanding environment provided by the nuclei of higher organisms. A possible explanation (consistent with our simplified discussion) is that palindromes were scarce when prebiotic self-reproducing strings appeared, because they could not replicate as easily.

2.2. Value and the "Order Parameter"

In studying self-replicating polymers one hopes to observe a tendency for the system to organize itself, selecting certain polymers to be more prevalent than others. This evolution from a uniform distribution of polymers to a distribution concentrated at a small set of polymers with high *selective value* should be observable in a Monte Carlo simulation, because under appropriate conditions we expect the uniform, unorganized state to be unstable with respect to small fluctuations. Generally the appearance of order in a system is associated with the introduction of a new thermodynamic variable, the *order* parameter.[1,13] Order can be said to emerge when a system displays a *broken symmetry*, i.e., a behavior that does not exhibit the full symmetry of the laws of motion of the system. In our case, the choice of a particular polymer using the cycle described above would be an example of a broken symmetry, since the chemical rules governing our simple model do not differentiate between nucleotide sequences. We note, however, that in fact the putative "symmetry" is, in the real world, a consequence also of nonequilibrium considerations, namely the complementary base pairing mechanism described above.

An order parameter generally expresses more than just the degree of order in a system[13]; it also mediates the coupling of the system to an associated external field. As an example of these two characteristics of an order parameter, consider a ferromagnetic material. The order parameter describing the transition to ferromagnetism is the local magnetization, $M(x, t)$, which is determined by the average of the spin magnetic moments in the neighborhood of x. If the moments are aligned, then M is nonzero, indicating an ordered state. A weak external H field applied to the ferromagnet will couple to the magnetization, selecting a preferred direction. The arbitrary choice of direction of M in the absence of an applied field represents the "broken" symmetry of the system, since the Hamiltonian for the system is rotationally invariant, unlike the chosen ground state. The ground state is degenerate under symmetry transformations of the Hamiltonian. If the order parameter η is a macroscopic continuous variable, its direction or *phase* must be continuous and slowly varying. To maintain this coherence, any change in $\eta(x)$ must cause $\eta(x')$ to respond. The transfer of a generalized force through an order parameter field is known as the property of *generalized rigidity*, describing the long-range coherence of an ordered system.

This spatial dependence will not concern us below, since our actual model describes a well mixed, homogeneous "test tube" full of polymers. The order that we require is an ordering not in physical space but in "value space". However, an amusing analogy can be drawn between generalized rigidity and living systems. Once an order parameter does emerge, it will coupled to the selective forces present in the environment, which will cause the order parameter to adjust to minimize these forces. The principle of generalized rigidity suggests that the order parameter will respond throughout the system to forces applied in one region. Boundary effects might yield a nonuniform "texture" with η varying throughout the region, but the important feature is that selective pressure in one area may affect all connected regions. It may not be completely unreasonable to suggest that the observed uniformity of the genetic code may be related to the similar properties of the prebiotic soup.

What order parameter should we use to describe these systems? The previous paragraphs suggest that the field which couples to the order parameter drives the system to a higher selective value, so it is natural to associate the value of information with some aspect of the order parameter. The initial degeneracy of our system with respect to particular sequences of monomers (assuming an original random distribution and kinetic equations that do not prefer any sequence) imparts a discrete, "phase"-like character to η since the choice of a sequence represents the breaking of the symmetry between different sequences. For strings of length N, the ordered state has an exponentially large degeneracy. Quastler[14] discusses this degeneracy in his consideration of value by noting that strings acquire value via selection (see also Eigen.[15,16] The amplification of small fluctuations drives the system away from the undifferentiated state into a condition where a polymer or group of polymers predominates.

Value is therefore imparted to a polynucleotide by its ability to survive and reproduce, which in turn is affected by local environmental conditions, such as the concentrations of other polynucleotides, polypeptides, ions, etc. These physical interactions define a value system that one can use to decide which polymers are most fit to replicate. To quantitatively study such systems, it is helpful to consider a simple linearized model of some phenomena which we expect to observe, as given in the next section.

2.3. A Simple Linear Model

For our purposes, the state of a homogeneous polymer system can be characterized by the number density x_j of each polymer type. The average time evolution of such a system is described by the (vector) differential equation

$$dx/dt = \mathbf{R}(\mathbf{x}, t) \tag{4}$$

where R will generally be a nonlinear function of its arguments. Since the growth of a particular species is determined primarily by a self-instructing step, we can write

$$\dot{x}_i = (r_i - d_i)x_i + e_{ij}x_j \tag{5}$$

where summation over repeated indices is implied. The r_i represent the state of production of string i from the template i, and the e_{ij} represent the production of string i via the imperfect replication of template j. The constants d_i take into account environmental factors which may cause the ith polymer to be effectively removed from the system via breakage, conformational changes, etc. The general time evolution for a given set of initial conditions $x_i(t = 0)$ is given by

$$x_i(t) = \sum_k b_k \xi_i^{(k)} e^{a_k t} \tag{6}$$

with the b_i given by the initial conditions via

$$x_i(t = 0) = \sum_k b_k \xi_i^{(k)} . \tag{7}$$

The $\xi_i^{(m)}$ are eigenvectors of the matrix $R_{ij} = (r_i - d_i)\delta_{ij} + e_{ij}$ and the a_m are their corresponding eigenvalues.

While this model ignores the important nonlinear effects generated by substrings combining to replicate a polymer, and assumes an unchanging supply of activated monomers, it illustrates some interesting phenomena associated with replication. Considering the behavior of the system at large times t, the largest eigenvalues a_0 dominates, and the densities vary as

$$x_i(t) \rightarrow \xi^{(0)} e^{a_0 t} b_0 . \tag{8}$$

The eigenvector(s) of the largest eigenvalue determine the eventual relative concentration of the polymers, since the growth of this "normal mode"

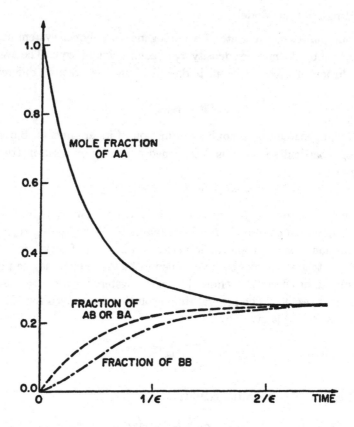

Fig. 2. Decay of an ordered system in the linear model of Sec. 2.3.

will eventually outstrip all others. An example of a solution of the linear equations in a very simple case is shown in Fig. 2.

To add selection to our model, we can add non-zero "death" probabilities d_i, or equivalently alter r_i. The problem is now reduced to finding the eigenvalues and eigenvectors of a nearly diagonal matrix, a problem that is easily soluble using perturbation theory, yielding eigenvectors

$$\xi'^{(n)} = \xi^{(n)} + \frac{e_{kn}}{r_k - d_k - r_n + d_n} \xi^{(k)} \tag{9}$$

with eigenvalues

$$a'_n = a_n + e_{nn} \equiv a_n . \tag{10}$$

If a_n is the largest eigenvalue then for long times only $\xi^{(n)}$ will survive, with small amounts of other polymers, having mole fractions on the order of $e_{kn}/(r_k - r_k + d_n - d_n)$ for the kth polymer. Eigen (1981) has referred to these peaked distributions as "quasispecies" — a primary polymer with a coterie of mutants associated with it. Quasispecies, the normal modes of the linearized system, can be said to compete against each other in the sense that the fittest quasispecies (i.e., largest eigenvalue) will survive if one begins with a mixture of random polymers.

We have considered here the continuous-time, averaged version of an inherently stochastic process, and our approximation masks certain interesting properties (Eigen, 1971). The evolution of the system can be considered as a random walk, with a finite chance that the walker will either reproduce or die at each site. By analyzing such a walk, one obtains probabilities that the system will die before exponential growth begins. Statistical fluctuations can also kill off one quasispecies, leaving a less fit quasispecies to dominate. This is an expression of Quastler's conception of information value arising from a "frozen accident".

What is the fate of two polymers whose $(r_i - d_i)$ values are nearly equal and large, but which can mutate into each other via n intermediate mutations? For example, the string AAA can mutate into BBB via the sequence of three mutations AAA→AAB→BAB→BBB. Especially if the intermediate sequences themselves have a large death probability, it will be improbable that the original string will mutate to the other one in a short time. We can model such a system by a three-state collection, where polymers 1 and 2 are stable and can both mutate into and from polymer 3, which has $(r_3 - d_3) = b$. The matrix R is given by

$$R_{ij} = \begin{bmatrix} 1 & 0 & \varepsilon \\ 0 & 1 & \varepsilon \\ \varepsilon & \varepsilon & b \end{bmatrix} \tag{11}$$

which has approximate eigenvalues $1 + 2\varepsilon^2/(1 - b)$, 1, and $b - 2\varepsilon^2/(1 - \beta)$ for $\varepsilon, \beta \ll 1$. The corresponding eigenvectors are, to first order,

$$\xi^{(1)} = \begin{bmatrix} \varepsilon \\ \varepsilon \\ 2\varepsilon^2/(1 - b) \end{bmatrix}, \quad \xi^{(2)} = \begin{bmatrix} 1 \\ -1 \\ 0 \end{bmatrix}, \quad \text{and} \quad \xi^{(3)} = \begin{bmatrix} \varepsilon \\ \varepsilon \\ 1 - b \end{bmatrix}. \tag{12}$$

Since two mutations are needed, $r \sim (1 - b)/\varepsilon^2$, a much slower relaxation time than for direct mutations. In general, the time constant for the decay of a single species to a collection of M other stable polymers, each separated from the other by n mutations, will vary as

$$r \sim (1 - b)^{n-1}/M\varepsilon^2 \tag{13}$$

where b is a representative stability of the intermediate strings, and ε is the error rate for single mutations. As n increases, the metastability of a particular polymer increases rapidly.

So far we have considered only self-instructed polymer replications, but many of these results carry over to complementary replications. The matrix R is now a block diagonal matrix with 2×2 blocks with diagonal entries $r, \bar{r}$, plus a small (off-diagonal) error matrix e'_{ij}. The eigenvectors of the system are

$$\xi^{(\pm)} = \begin{bmatrix} \sqrt{r} \\ \pm\sqrt{\bar{r}} \end{bmatrix} \tag{14}$$

(with eigenvalues $\pm\sqrt{r\bar{r}}$,) so that the ratio of the complement concentrations as $t \to \infty$ will be the square root of the ratio of the two reaction rates, $(r/\bar{r})^{1/2}$.

2.4. Stability and the Death Function

As we have seen in the previous section, a particular polymer can only be selected when the value of $(r_i - d_i)$ differs from polymer to polymer, a quantity which Eigen et al.[15,16] refer to as the "selective value" of the string. As a system evolves, the average value of this parameter increases, demonstrating a tendency towards self-organization. Note that while the value of the strings increases on average, the information content decreases, since a few messages will dominate the population. Many very complex factors influence the magnitude of d_i, such as interactions with neighboring or bonded polypeptides, inorganic catalysts, breakage, etc. Despite the difficulty of computing the d_i, there are some general properties which we can deduce from the existence of a distribution of nonzero d_i.

While d_i is a probability per unit time (a rate), we can in principle study the effects of various external influences by considering the extra energy that these forces contribute to the polymer; whether they be stabilizing or destabilizing influences. We write this interaction energy $D_N(\mathbf{S})$, where $\mathbf{S}$ represents a polymer with elements S_i. A large value of this *death function* indicates a destabilizing influence of the environment and a larger probability of "dying", or not belong able to reproduce. To convert from

$D_N(S^{(i)})$ to d_i, we need only use a monotonic map from the possible values of D into the unit interval. The details of our discussion and simulation will be slightly altered by the particular choice of this function, but all such monotonic maps surely belong to the same universality class. Since the $D_N(\mathbf{S})$ are essentially the energies of the various conformations with respect to some zero of energy, we might, for example, use the function

$$P_D = \frac{e^{D_N + \mu(N)}}{1 + e^{D_N + \mu(N)}} \qquad (15)$$

in analogy with the Fermi-Dirac distribution for a two-state system. This analogy is only a heuristic one, since the Fermi-Dirac distribution gives probabilities for equilibrium systems, not rates, yet this conversion illustrates a plausible relationship between $D_N(S)$ and d_i. Note that we have included a "chemical potential" μ which determines the overall death rate.

Given a collection of P_D, one can imagine a set of surfaces in "information space" assigning a measure of the reaction of the system (i.e., $D_N(S)$) to each individual polymer, i.e., defining the value in the sense of Volkenstein. Peaks in P_D will be unstable points, since these polymers will be more likely to die out. Troughs or minima, on the other hand, will survive, so that most of the remaining polymers in the system will live near these minima. If two minima are adjacent to each other, or the activation energy of the transition is not too great, these two polymers will form their own "pseudospecies", differing from quasispecies in that they are not dominated by one polymer, but by two.

The time evolution of the system is now describable by a random walk through the lattice with a finite death and replication probability at each site. As discussed previously, to pass through a hump in the death distribution requires a time on the order of $\tau \sim D^{n-1}/\xi^n$ where ξ is the mutation rate and D represents a characteristic height of the barrier; conversely, the rate of passing over a maximum is $1/\tau$. A state will be localized around a minimum if this rate is small.

To model the many varied factors that determine D_N, we assume[3] that due to the interplay of many unrelated and incommensurate effects, D_N will be a nearly random function of the S_i. In a system with only two bases, we can represent S_i as a "spin" with values $A = +1$, $B = -1$, so that any function of $\mathbf{S}$ can be expanded as a linear function because $S_i^2 = 1$ and $S_i^{2j+1} = S_i$. We write

$$D_N(S) = h_i S_i + J_{ij} S_i S_j + c_{ijk} S_i S_j S_k + \cdots . \qquad (16)$$

If we let **S** and its conjugate $-$**S** behave identically with respect to the influences that determine D_N, assuming detailed balance, then $D_N(\mathbf{S}) = D_N(-\mathbf{S})$ and

$$h_i S_i + J_{ij} S_i S_j - \cdots = -h_i S_i + J_{ij} S_i S_j + \cdots \qquad (17)$$

and all products of odd numbered collections of S_i must vanish ($h_i = c_{ijk} = \cdots = 0$). To third order, we are left with

$$D_N(\mathbf{S}) = \sum_{ij} J_{ij} S_i S_j . \qquad (18)$$

How many stable polymers are there with such a death function? To make progress in answering this question, we will consider a special choice for the J_{ij}. Owing to the enormous complexity of interactions, the net effect of the environment is fixed, but varies dramatically from polymer to polymer, so we can simplify the expression for $D_N(\mathbf{S})$ by taking J_{ij} to be $J\sigma_{ij}$, where J is a suitable coupling constant and σ_{ij} is randomly ± 1. We will use the same J for all N, since each additional monomer adds N terms of the form $J_{ij} S_i S_j$ which are essentially random.

To determine the number of stable molecules, we first calculate an approximate "density of states" $\rho(D_N)$, the fraction of polymers having a given death function. Since D is the sum of $N(N-1)/2$ numbers, each of which is $\pm J$, we can approximate the result by a random walk in one dimension with step size J and $N(N-1)/2$ steps, yielding a Gaussian distribution of standard deviation $J(N(N-1)/2)^{1/2}$ and a mean of zero. For large N, the Gaussian approximation is very good (Fig. 3). For a Gaussian distribution this will be about 16% of all polymers. As N gets large, more and more branches are added to the tree described earlier; at each N there are $N \times 2^{N-1}$ branches, so that we expect long strings to form many pseudospecies, and these groups rather than any one species will be selected.

Our simple random walk computation ignores some important properties of the function $\sum J_{ij} S_i S_j$. By analogy with a spin glass, which has the same type of Hamiltonian, we expect the polymers to be "frustrated" in their attempts to get a very low value of D_N, since it is impossible to add all of the terms $J_{ij} S_i S_j$ with the same sign. If we try to align S_i and S_j so as to have $J_{ij} S_i S_j$ positive, we will inevitably make some other $J_{ik} S_i S_k$ negative. Such an arrangement leads to many dead ends in the search for an

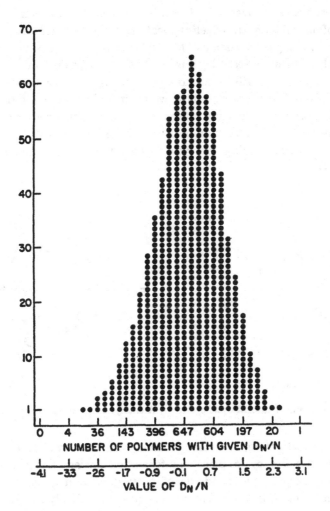

Fig. 3. The number of polymers of length N having death function D_N/N. For large N, it is well approximated by a Gaussian.

absolute minimum, and produces many local minima. These local minima build in a diversity of possible strings to be selected.

2.5. Coevolution and Hypercycles

Thus far we have assumed that selection occurred among polynucleotide information carriers first, and then among polypeptides. We have therefore chosen between the proverbial "chicken and the egg" of molecular biology, because of our requirement of self-replication. Eigen[15,16] has taken another approach, avoiding the question of the first selected molecules by beginning with a collection of cycles. Arguing that all polypeptides have some catalytic function, there will be random cooperative behavior between early polynucleotides and proteins, which would in some instance form closed loops, with each molecule being catalyzed and catalyzing another molecule. These *hypercycles*, which contain both protein and nuclei acid components, can evolve by restructuring the cycle to include more reproducible parts (see Fig. 4).

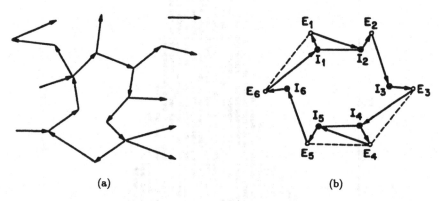

(a)   (b)

Fig. 4. (a) A single autocatalytic cycle. (b) A catalytic hypercycle (after Eigen).

As mentioned before, the main problem with such an approach is the restrictions that such cycles place on information carriers. Current polynucleotide polymerases are hundreds of thousands of daltons, and it is therefore difficult to imagine any prebiological (i.e., nearly random) peptide being very efficient in catalyzing replication, and vice versa. Since nucleotide molecules can also possess tertiary structure by virtue of their complementary binding, it is more likely that at first nuclei acids were the only evolving part of the prebiological soup. Alternatively, the two polymer systems may have evolved separately, the proteins perhaps following a non-specific route of the sort envisaged by Dyson,[17] until they reached a level at which they could help each other in a symbiotic relationship. A scheme suggested by Kuhn and Wasser[18] allows for the simultaneous development of complex proteins and nucleic acids, using a proposed aggregation property of palindromic RNA's to limit the "genetic drift" of a species. Errors will result in RNA's which cannot aggregate, so only correct copies will survive. This program does not deal with the tendency of partially denatured RNA's to renature without combining with other strings to replicate, which would result in a much reduced growth rate for palindromes, as described above. The chaotic death function described in this chapter gives RNA strings the required stability as demonstrated below, without relying on this specific and essentially untestable hypothesis. A large portion of the protein replicating machinery consists of nucleic acids, including ribosomal RNA (rRNA) and the transfer RNA's which direct the decoding of RNA strands. The conformation of simple, self-complementary strands like tRNA could have allowed them to perform rudimentary enzymic tasks. Note that a $D_N(S_i; [S_1], [S_2], \ldots [S_p])$ which depended on the concentrations of other polymers would simulate this effect and possibly mimic a hypercycle.

3. Some Numerical Results

The model described above has been implemented as a computer simulation, whose results are described below. We have essentially simulated the cycle of Fig. 1 using a few lumped parameters to represent the various rate constants for replication and death.

3.1. Defining the Model

We begin the simulation with a small number of dimers and trimers to act as templates for replication. This "soup" of small polymers contains equal numbers of each type of dimer and trimer to preclude any bias towards particular strings which might later appear to have been selected. At the beginning of each cycle (or *generation*), a flux of both A and B monomers is added to provide raw materials for growth and replication. From this collection of polymers, one is chosen as a template, and is compared with the remaining polymers one by one to determine if they are sufficiently complementary to form a double stranded complex. With a probability dependent upon the degree of complementarity, the two polymers stick together. Another possible complement is then selected from the soup, which is tested for complementarity with the part of the template which is not already matched with the first complement. This procedure continues until a certain number of consecutive unsuccessful attempts have been made to find a complement, or until the template has been completely matched.

The (template + complements) complex is then set inside, and a new template is chosen from the remaining single stranded polymers. This process repeats until all the strings have been used or until a few templates in a row have not found any complements, and corresponds to the renaturation in Fig. 1.

To determine whether or not a string is sufficiently complementary to the template to form a "double helix", the two strings are placed parallel to each other, shifted by a random amount. Each weak complementary bond which must be formed is assigned a probability, $p(\text{A-A})$, $p(\text{A-B})$, and $p(\text{B-B})$, which describes the accuracy of recognition. For the probability that the two strings will match, we use the product of the probabilities for each weak bond. An error corresponds to a nonzero $p(\text{A-A})$ and/or $p(\text{B-B})$. When the complementary string hangs over the end of the template, i.e., some monomers in the complement are not matched with monomers in the template, a factor of $p(\text{A-end})$ or $p(\text{B-end})$ is included for each unmatched monomer. Physically, this accounts crudely for the different behavior of the ends of polynucleotides, which can be bound to ions, proteins, etc., that can interfere with the binding of complement to template. Practically, $p(x\text{-end})$ controls the lengthening rate of polymers; if $p(x\text{-end})$ is zero, the complementary copy cannot be longer than the template, and no growth can occur. To decide whether or not a complement will bond to a template, a computer generated random number uniformly distributed between zero and one is compared with the total probability of bonding; if the random number is less than this probability, the polymers form a double stranded complex.

After the "renaturation" part of the cycle each complex is examined for juxtaposed complements; these can then form strong covalent bonds to combine two short strings into a longer one. This process is governed by the probabilities $p(\text{A} = \text{A})$, $p(\text{A} = \text{B})$ and $p(\text{B} = \text{B})$. For example, the complex

```
complement   A=A=A=B=B=A  B=A=B=A     A=B=A
             | | | | | | | | | |     | | |
template     B=B=B=A=B=A=B=A=B=A=B=A=A=A=B
```

which contains an "error" at the third site of the template, can form a covalent bond between the sixth and seventh monomers on the "complement" side. The complement of length three cannot be joined to the others, since it is not adjacent to another complement. Note that for small double-bonding probabilities, more subreplicas of the template are produced, while for large probabilities a more complete complementary copy can be made. After the covalent bonds have been formed, the weak bonds are broken (the "denaturation" part of the cycle) and we are left with a new soup of single stranded polymers.

The application of the "death function" occurs immediately after the denaturation step; the death function $D_N(S)$ and the corresponding death probability are computed for each polymer as it is peeled from its template-complement complex. Again, a "random" number between 0 and 1 is generated to determine the fate of the polymer. The "death" of an RNA strand can be envisioned in many ways — decay into monomers, breakage, reproductive "death" in which the polymer is chemically altered to prohibit future pairing, or simply the removal of the polymer from the soup, perhaps by adsorption onto a rock, etc. To prevent a proliferation of polymers which would rapidly exceed the computer memory, we choose the latter, and discard all "dead" strings. Once all paired complexes are denatured and checked for survival, the new soup of single strands is used to repeat the cycle.

The death probability is given by the Fermi-Dirac-like law

$$P_D(\mathbf{S}) = \frac{\exp(J(d(\mathbf{S}) + \mu))}{1 + \exp(J(d(\mathbf{S}) + \mu))} \tag{19}$$

where $d(\mathbf{S})$ is the "reduced death function" obtained by normalizing the distribution of $D_N(S)$ to a Gaussian of unit standard derivation

$$d(\mathbf{S}) = \frac{D_N(\mathbf{S})}{J\sqrt{N(N-1)/2}} . \tag{20}$$

This death process is controlled by the two parameters J and μ. The "chemical potential" μ prescribes the value of the reduced death function at which a polymer has a 50% chance of survival. The coupling constant J behaves like an inverse temperature, and describes the severity of the death function. For large J (zero temperature) the cutoff at μ is sharp, and all polymers $\mathbf{S}$ such that $d(S) < \mu$ survive while all others die. Reducing the coupling constant results in a more forgiving environment where a low value of the death function becomes less important for survival (Fig. 5).

3.2. Complementary vs. Self-Replicating Kinetics

In all results reported below, we have maintained the complete dynamical symmetry between the two types of monomers by choosing $p(\text{A-A}) =$

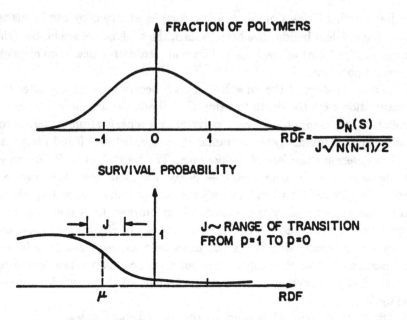

Fig. 5. (Top) A schematic plot of the "density of states" function also represented in Fig. 2. (Bottom) The survival probability $d(S)$ as a function of the reduced death function. Note that the cutoff depends on μ and its width on J.

p(B-B) and p(A = A) = p(B = B) (recall that $D_N(S) = D_N(-S)$ by construction, so death is symmetric). If we begin with an initial soup which also preserves this symmetry, one expects that two complementary strings S and $-$S will be produced at an identical rate, so their concentrations should be equal. For computer simulations with complementary kinetics, i.e., p(A-A) small and p(A-B) unity, this is indeed the case, with S and $-$S appearing equally to well within $\pm\sqrt{N}$, as expected for counting errors. Adding a death function with various μ and J does not alter this equivalence.

In order to reproduce itself, a string S must first make a complementary copy $-$S (or something resembling $-$S) to be used as a template for production of more S. Under optimal conditions of perfect recognition (p(A-A) $= 0$, p(A-B) = 1) and perfect polymerization (p(A = A = p(B = B) = 1) this process can take two generations. Allowing errors ("mutations") will slow this process down, as will the possibility of producing S by linking fragments made by other polymers. An excess of S over $-$S will result in the production of more −S, tending to restore the equality of the concentrations of the two polymers. There is a built-in resistance to breaking the symmetry between S and −S.

Having demonstrated that our computer simulation with complementary pairing (like that observed in nature) does preserve the equality of S and −S, even for error rates (p(A-A) = p(B-B)) on the order of 10–15%, it is convenient to eliminate the intermediate step of producing a complementary copy in order to reproduce S. This is useful for two reasons: First, the growth rate is approximately doubled, and second, the analysis is simplified by eliminating the need for S and −S to be considered part of the same "species", as defined below. Note that the restoring force of complementary kinetics is now absent, since S and −S are no longer dependent on each other for reproduction. They behave as independent polymers, competing for the same raw materials, and need not be present in the same quantities. As will be seen below, their concentrations are in fact rarely comparable, with one of the two dominating if either of them exist. For the intermediate case of random reproduction (no recognition, any string can be a template for any other) the concentrations of S and −S are completely random and fluctuate in time.

In what follows, we will always be considering self-replicating kinetics unless otherwise specified, concentrating on the competition between polymers that are not mirror images of each other. It is the possibility of selection between these polymers that interests us as a model for the selection of prebiological information, so we choose the simplest member of the universality class of these models to study. Note that comparisons with complementary kinetics data have shown that self-replication runs show the same properties as complementary runs with the replacement of a pair (S, −S) by (S) or (−S). Furthermore, the symmetry of S and −S (which led to our choice of $D_N(S)$) is still preserved.

3.3. Life Without Death: The Absence of Value

Even in the absence of a death function, one cannot say *a priori* what results the above model will yield. There are complicated nonlinearities which distinguish our model from the simple linear model (Sec. 2.3) à la Eigen. A template will often only replicate fragments of itself; these fragments can in turn reproduce wholly or in part, they can be joined together to recreate the original template, or they might join in a different order with other polymers to form a string completely different from the original. There is a great deal of competition for small fragments, which have not yet specialized and can be used to replicate many different templates; longer strings are more restricted in their choice of a template, but can also serve as templates themselves.

The amount of "food" (small polymers) is limited, and the larger polymers which use them first by chance will have gained a competitive advantage over the others. The "frozen accident" would be completely random, and one would be hard pressed to identify it with biological selection. We have observed that in simulations without a death function and without mutation, the distribution of polymer concentrations is biased towards higher concentrations than would be expected by a Poisson distribution — too many polymers have more than their share of the soup. Note that the population is not dominated by a handful of types, so the "selection" is not striking. Which polymers are successful varies from run to run, and appears to be random. The addition of small-medium mutation probabilities ($\simeq$ 10–15%) does not destroy this effect but does diminish it, and as the mutation and recognition probabilities approach each other we return to a random collection of polymers. Adding mutation allows drifts which cause the successful polymers to change in time.

In an attempt to characterize the polymers favored by this process, we calculate the "triplet parameter" used by Anderson and Stein.[1] Given a string of monomers, we can describe it by specifying the lengths of the alternating strings that make up the polymer. Each alternating sequence ends at a repeated monomer, AA or BB. For example, the string

$$ABABAABBAAB = (ABABA)\,(AB)\,(BA)\,(AB)$$

can be written as a sequence of alternating strings of lengths (5,2,2,2), and contains the triplets (5,2,2) and (2,2,2). Note that the sequence of alternating string lengths, plus the first monomer of the string, uniquely defines the entire polymer; for example, (1,1,2) can only refer to AAAB or BBBA. This implies that for a random uncorrelated collection of polymers the two triplets (a, b, c) and (a', b', c') will have the same probability if their lengths $(a + b + c)$, $(a' + b' + c')$ are equal, and they should therefore have the same occurrence rate in a random sequence of A's, B's. Including a simple model for the distribution of lengths of polymers (since triplets can also start and end at the ends of polymers), we again get dependence only on $(a + b + c)$. This model agrees well with the random recognition runs.

For small mutation rate, the occurrences of triplets depend very strongly on the relative values of a, b, and c, not just on their sum. This occurrence rate is, however, independent of permutations of (a, b, c). For example, (1,1,3), (1,3,1) and (3,1,1) all have similar occurrence rates, which are equal up to the statistical error incurred by the Monte Carlo simulation, but they differ greatly from the permutations of (1,2,2); the extent of this discrepancy depends upon the error rate (see Fig. 6). This apparent intrinsic ordering in the system can be considered an "emergent property". Long runs to determine the stability of this effect have not yet been made.

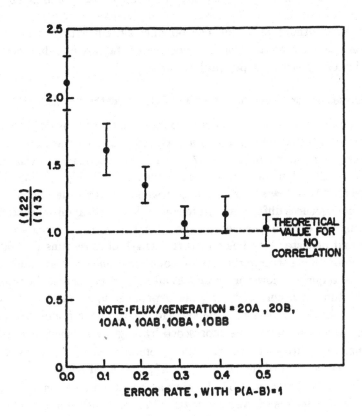

Fig. 6. With small error rate, different configurations of the same length have different survival behavior with $D = 0$. As error rate increases, this effect is washed out.

One possible explanation for this correlation is that (113) used 4 of one monomer and 1 of the other to replicate, while (122) uses them in a 3:2

ratio. Since monomers are present in equal concentrations in the soup, the (122) sequence will be able to make more copies, and will dominate, the other being limited by the small flux of monomers at each generation. In general, (a, b, c) will be preferred if a, b and c are predominantly even and suppressed for a, b or c odd, with the effect being most important for small $a + b + c$. While a random collection of A's and B's, present with equal probability, would have (113) equals to (122), the manner in which the strings are constructed has imposed additional structure on their distribution.

Without a death function, the soup remains primarily undifferentiated, growing quickly but not dominated by any sequence or small group of sequences. Without providing our polymer system with a value (death function), we see that the desired properties of stability and differentiation (selection of a *few* strings per run) are absent.

3.4. Speciation, or "Into the Valley(s) of (the) Death (Function)"

The addition of a death function to our procedure provides the model with growth, differentiation, and stability. The cases we investigated most thoroughly correspond to $\mu \sim -0.7$ to -1.0 and various J, so that about 10–15% of all polymers survive (see shaded area in Fig. 7). Each run is dominated by a few species after approximately 100 generations. These dominant sequences differ from run to run under identical initial conditions, exhibiting the diversity needed for life. Furthermore, the sequences were stable, and persisted (see below) for the length of most runs ($\sim$ 500–600 generations). If we observe the relative concentration (mole fraction) of any individual sequence, however, in almost all cases it peaks and then decays, with the time of occurrence the peak dependent upon the length of the sequence. As time progresses, the average length of the population grows (see Fig. 8) and sequences are superseded by longer versions of themselves. It is therefore natural to consider polymers with similar sequences and subsequences as part of the same "species", as defined below.

A species is constructed by choosing a generation and considering the sequences of polymers of average length. The soup is then searched for all smaller strings ("ancestors") and larger strings ("children") whose sequences match wherever comparison is possible. Small (length ≤ 5) polymers are not included in the species, since these short polymers are ancestors to many species and can be considered to be undifferentiated

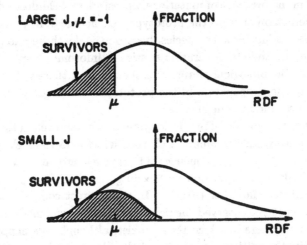

Fig. 7. The sharpness of the cutoff between surviving species and non-surviving ones depends on the magnitude of J, which acts as an inverse "temperature". The graph shown is the product of the density of states with the survival probability.

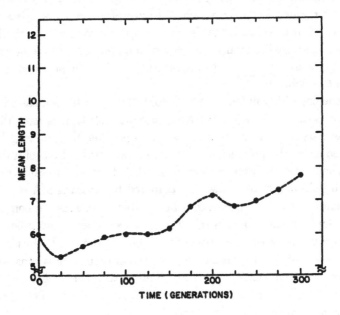

Fig. 8. As time increases, the mean length of surviving polymer species also increases.

materials. In the presence of mutations, ancestors and children which differ only by one monomer from the prototype are included. In practice, there is very little overlap between species, because the death function prohibits many sequences, thereby separating species in information space (which is quite large). The biological picture of a species is all those polymers which contribute to the propagation of the prototype. Each valley of the death function corresponds to one species.

The diversity required of prebiotic systems is manifested in the selection of a few species which dominate each run; which species is chosen varies from run to run. When two sequences of the same length and death function are placed in direct competition by "seeding" the initial soup, one species rapidly dominates the other (Fig. 9). In these direct competition simulations we have not yet observed the development of a symbiotic relationship between the polymers in which they coexist, although this happens often when the initial conditions are not "seeded". Under seeded conditions, the survival of a sequence depends strongly on the structure of that particular valley of $d(\mathbf{S})$ which lies below $\sim \mu$. If the valley is broad, the seeded polymer will immediately produce many ancestors giving it an advantage over the narrower valley sequence which can only produce a few types of ancestors. Without seeding, longer length ancestors are produced slowly, via undifferentiated small polymers which are common ancestors to all species. It is plausible that such a high concentration of small polymers is more favorable to symbiosis.

Lengthening of a polymer sequence corresponds to the increasing specificity of an organism — while there are more ancestors, it is more difficult to put them together in the correct way. It is possible that, on lengthening a given sequence, the polymer produced will have crawled out of its valley, i.e., $d(\mathbf{S})$ may then be greater than $\sim \mu$. The tendency towards increasing length (which can, of course, be tempered by lowering $p(A = A)$, etc. and $p(x\text{-end})$) pressures species into being able to increase in complexity and "develop" to keep up with the others. Theoretically, one valley could branch into two species, each better adapted to some niche in the environment (valley of $d(\mathbf{S})$). Presumably this would require runs considerably longer than we have made, which are limited to mean lengths of 15–20 monomers. Mutation rates $\leq 15\%$ do not destroy the effects of competition and the subsequent selection of species.

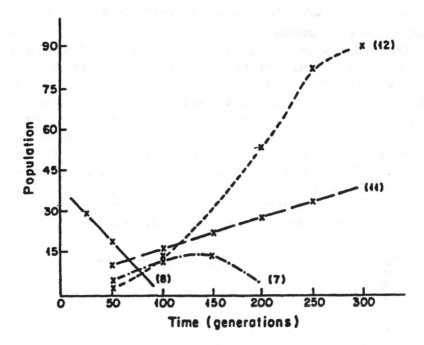

Fig. 9. A sample run, showing the fates of different selected related species. The numbers in parentheses represent the lengths of the strands chosen.

3.5. Evolution of Adaption

The previous section showed that "competition for nutrients" can lead to the coexistence of a few species. The representative sequence of each species steadily lengthens in time as the species searches for the optimal sequence for the given environmental conditions (i.e., given J_{ij}). This increase in complexity to make the best use of the environment is one prerequisite for evolution. A changing environment requires a population which can adapt to new conditions. This variation of the environment was simulated by allowing the σ_{ij} to vary in time (recall $J_{ij} = J\sigma_{ij}$). Two random sets $\sigma_{ij}^{(1)}$ and $\sigma_{ij}^{(2)}$ were chosen, and $\sigma_{ij}(t) = a(t)\sigma_{ij}^{(1)} + (1 - a(t))\sigma_{ij}^{(2)}$, where

$$a(t) = \begin{bmatrix} 1 & t < t_c \\ 1 - (t - t_c)/t_v & t_c < t < t_c + t_v \\ 0 & t_c + t_v < t \end{bmatrix}. \qquad (21)$$

Periodic $a(t)$ have also been used, in which $\sigma_{ij}(t)$ oscillates between $\sigma_{ij}^{(1)}$ and $\sigma_{ij}^{(2)}$ every few hundred generations.

One result of a time varying σ_{ij} is shown in Fig. 10, with a mutation rate (see below). As usual, initial growth under $\sigma^{(1)}$ leads to the selection of a few species (A,C). As the environment changes to $\sigma^{(2)}$, one of the species cannot survive (A), since this minimum of $d^{(1)}(\mathbf{S})$ does not correspond to a minimum of $d^{(2)}(\mathbf{S})$. Another species (B), present in trace amounts before the change, does coincide with a minimum of $d^{(2)}$, and dominates after $t_c + t_v$. The only example of a true adaptation, however, is species B, which suffers initially, but then mutates to align itself with a more favorable valley of $d^{(2)}(\mathbf{S})$. This alteration of B can be observed by studying the individual sequences in detail. The mutation allows the species to adapt. Note that "species" here means the same sequence to within one error.

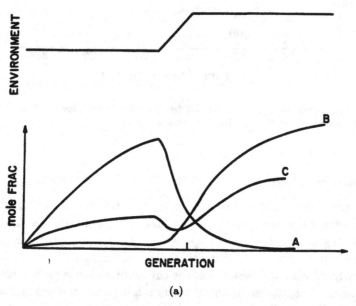

(a)

Fig. 10. (a) Adaptation. The environment (i.e., the J_{ij}) was changed during the course of 100 generations. In this time, species A, which had been doing well, died out; species B, which had been doing poorly, began to grow; and species C began to die out but then "mutated" (i.e., a replication error was made) and its antecedents thrive. (b) Changing environment. The dashes at 200, 350 and 500 generations represent environmental changes. Polymer 1 responds to first change and thrives. Polymer 2 could not adapt, and dies off. For this run $J=\pm 2.0, \mu=-0.7$, and the error rate was 2.5%. Error bars indicate noise.

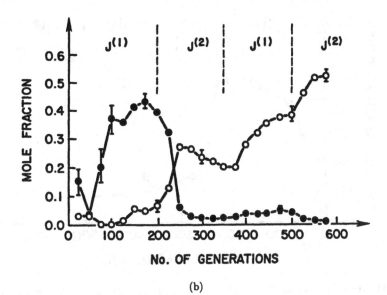

(b)

Fig. 10. (*Continued*)

Periodic environment changes also lead to interesting phenomena (Fig. 11). Beginning at generation 200, $\sigma(t)$ alternated between $\sigma^{(1)}$ and $\sigma^{(2)}$, changing every 150 generations. Species 2 thrived on $\sigma^{(1)}$ but died when $\sigma^{(2)}$ took over. Polymer (1) grows modestly on $\sigma^{(1)}$ and takes over on $\sigma^{(2)}$. When $\sigma^{(1)}$ conditions are restored, species (2), which we know can dominate under these conditions, is not restored to its previous concentration. Species (1), having been in the very favorable environment $\sigma^{(2)}$, can now grow in $\sigma^{(1)}$, in which it was not overly successful in the previous cycle. Being suited to growth in $\sigma^{(1)}$ and $\sigma^{(2)}$, it continues to grow in the periodic conditions. The sharpness of the transitions from $\sigma^{(1)}$ and $\sigma^{(2)}$, modeled by t_v, does not qualitatively change this behavior, although a sharp transition (small t_v) does cause a more severe depletion of the soup, which "prunes" the family trees of the species and lets them grow anew. Small t_v allow each species to start off relatively even, while a large t_v favors those species which can be assembled from the family trees of other species not able to survive in $\sigma^{(2)}$. The growth of these species on $\sigma^{(1)}$ provides raw materials to fuel new species when the change comes.

Not only do certain species adapt to the new conditions, but in some

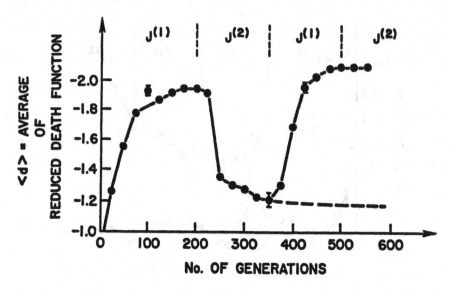

Fig. 11. Hysteresis. As in Fig. 10, the environment suddenly changes at $t = 200$, 350 and 500. Here we plot the average value of the original death function for the entire soup. The polymers remember that they had at one time grown in this environment, even when it no longer feels these J_{ij}.

sense the entire population "remembers" that it came from $\sigma^{(1)}$ conditions, and can therefore adapt faster when $\sigma^{(1)}$ environment returns. This is demonstrated in Fig. 11. The ordinate is the average reduced death function $d^{(1)}(S)$ for the soup, using $\sigma_{ij}^{(1)}$. In the $\sigma^{(2)}$ time periods, one would expect the totally independent death function $d^{(2)}(S)$ to govern the population, so that $d^{(1)} \to 0$. This is not the case; the population remembers its past environment, and adapts to it even better when it returns (compare 550 generations with 250). Changing initial conditions have in this case shaken up the soup, so that it lands in a better minimum than the one previously selected.

The "hysteresis", or dependence on past conditions, is reminiscent of the human appendix — future selection does not always eliminate vestigial features. It is interesting to consider this data from the viewpoint of "neural network" models along the lines of Hopfield's model.[20] The population has been "taught" to survive in $\sigma^{(1)}$, and when it "learns" to survive in $\sigma^{(2)}$ it does not "forget" its training. Unlike Hopfield-like models, however, "genetic" memory is not stored in the J_{ij} but in the state of the soup itself.

4. Conclusions

We have demonstrated that the model described above, a self-replicating collection of polymers whose value or viability in a given environment is described by a chaotic frustrated "death function", can exhibit the required properties of growth, differentiation (diversity), and adapation. These properties are relatively independent of the parameters of the model — mutation rate less than or comparable to 15%, survival rate roughly 10–20% (set by μ, J), over a wide range of lengthening conditions. These bounds represent only the regions investigated, not actual limits. Furthermore, the absence of value does not lead to all of these properties. The magnitude of our parameter $\langle d(S) \rangle$ is directly related to value (in some sense the average value of the system). Like the magnitude of an order parameter, it describes the organization of the system but not which species are chosen; these are the arbitrary phases of the order parameter. External fluxes of large polymers which seed the soup act as the generalized forces in this problem, since they can influence the "phase"-like aspects of the order parameter.

Present results indicate that the above model is a reasonable scenario for the emergence of valuable information in a prebiological context. In our model we have fixed the definition of value by choosing the J_{ij} in advance, or allowing them a simple variation, in order to examine more carefully the effects of a clearly defined value and how this value can cause a collection of polymers to order themselves. As mentioned above, value is only determined with respect to a trigger system, in our case the nucleic acid soup itself which can be triggered to select a polymer species. A more complex and perhaps realistic model would allow the soup itself to define (Anderson and Stein) value, and later to account for the influence of polypeptides, which are not required for our model. A consistent picture seems to be that the first self-organizing prebiological systems were made of nucleic acids, and that protein systems developed later or in parallel, and merged at some later stage.

Author's Note

After the initial work was completed, Dr. Lloyd Demetrius called our attention to the paper on de novo replication of RNA by Biebricher, Eigen and Luce.[21] Although the mechanism is different, the essential characteristics of our model seem to be embodied in an actual biochemical experiment. We consider it remarkable verification of our results that in essence the nature of their resulting RNA polymers is essentially identical with ours. Specifically, the outcomes of repeated experiments are similar but not identical and they have the same tendency for a few quasispecies to develop. Thus, in essence their experiments are a confirmation of our model's biological relevance, and at the same time a beautiful model of the beginnings of evolution.

Acknowledgements

This manuscript summarizes the results of an extremely enjoyable collaboration with D. Stein and P. W. Anderson. It is a pleasure to acknowledge countless conversations with A. Klausner, C. Thompson, T. Searle, M. Mayer, M. Armstrong, S. Loebl, M. Searle, and T. Slesar.

References

1. P. W. Anderson and D. L. Stein, "Broken Symmetries, Dissipative Systems, Emergent Properties and Life", in *Self-Organizing Systems*, ed. F. Yates (Plenum, 1985).
2. A. Rich, "Origins and Exobiology", in *Irreversible Thermodynamics and Origins of Life*, ed. A. Katchalsky (Gordon and Breach, 1974).
3. P. W. Anderson, "Suggested Model for Prebiotic Evolution: The Use of Chaos", *Proc. Nat'l. Acad. Sci.* **80** (1983) 3368–3390.
4. J. L. Fox, "Origins", in *Molecular Evolution Prebiological and Biological*, ed. D. Rohlfing and A. I. Oparin (Plenum, 1972).
5. E. Schrödinger, "What is Life?" (1944) from *What is Life and Other Scientific Essays* (Doubleday, 1956).
6. L. Brillouin, *Science and Information Theory (2nd ed.)* (Academic Press, 1962).
7. A. I. Khinchin, *Mathematical Foundations of Information Theory* (Dover, 1957).
8. M. V. Volkenstein, "The Amount of Value of Information in Biology", *Foundations of Physics* **7**, (1977), No. 1/2.
9. G. S. Rao., Z. Hamid, and J. S. Rao, "Information Content of DNA and Evolution", *J. Theor. Biol.* **81** (1979) 803–807.
10. H. F. Blum, "On the Origin and Evolution of Living Machines", *Am. Scientist* **49** (1962) 474–501.
11. B. Lewin, *Gene Expression I* (Wiley, 1974).
12. A. Lehninger, *Biochemistry (2nd ed.)* (Worth Publishers, 1975).
13. H. Haken, "Cooperative Phenomena in Systems Far From Equilibrium", *Rev. Mod. Phys.* **47** (1975) 67–121.
14. H. Quastler, *Emergence of Biological Information* (Yale University Press, 1964).

15. M. Eigen, "Self-Organization of Matter and the Evolution of Biological Macromolecules", *Naturwissenschaften* **58** (1971) 465.
16. M. Eigen, "Molecular Self-Organization and the Early Stages of Evolution", *Quart. Rev. Biophys.* **4** (1971).
17. F. J. Dyson, "A Model for the Origin of Life", *J. Mol. Evo.* **18** (1982) 344; *Origins of Life* (Cambridge University Press, 1975).
18. H. Kuhn and J. Wasser, *Nature* **298**, (1982) 585–586; *Experientia* **39** (1983) 834–841.
19. M. Eigen, W. Gardiner, P. Schuster, and R. Winckler-Oswatitch, "Origin of Genetic Information", *Scientific American* **244** (1981) 88–118.
20. J. J. Hopfield, *Proc. Nat'l Acad. Sci.* **79** (1982) 2554–2558.
21. C. K. Biebricher, M. Eigen, and R. Luce, *J. Mol. Biol.* **148** (1981) 369, 391.

Evolution of Species and Punctuated Equilibria: Genotypes, Phenotypes and Population Dynamics

Gérard Weisbuch

Groupe de Physique des Solides de l' ENS, 24 rue Lhomond,
F-75x231 Paris Cedex 05, France

Introduction

The evolution of species can be seen as a problem in dynamics. Its complexity results from the fact that it involves three levels: Genomes, organisms and populations. Furthermore the interrelations between the three levels are not fully understood. The purpose of this chapter is to show that simple guesses about the nature of these relations allow to predict generic properties of population dynamics. In particular this enables us to propose a solution to the apparent paradox of punctuated equilibria posed by paleontologists.

Let us consider the set of all possible organisms with a genome of a given length. If. e.g., the DNA length is N nucleotides, since 4 different nucleotides exist in DNA, the phase space contains 4^N different organisms. A population function is defined for each organism, and the time evolution of this function obeys a dynamics that we would like to describe by some differential system similar to Fokker-Planck equations. Variation of the population of each organism would imply internal terms depending on the biological properties of the organism (e.g., food absorption, reproduction and death) and coupling terms due to mutations. This nice program of theoretical investigations is blocked at its first stage because we have very little information about the way the biological properties of an organism — the so-called phenotype — are related to its genome.

The Genotype/Phenotype Problem

Although the central dogma of molecular gentics is that the phenotypic properties are determined by the genotype, the set of genes present

141

in the chromosomes, this relationship is far from being understood. The
first successes concerning the genetic code, i.e., the correspondence between
the nucleotides of the DNA and the amino acids of the proteins have not
been followed by similar progresses at higher levels: Even the cellular level
is little understood, not to mention individual organisms or populations.
Ontogenesis, the science to development of organisms from their first cell,
which itself depends on its genome, is presently *terra incognita*.

The following scheme of causalities although well admitted, is poorly
understood.

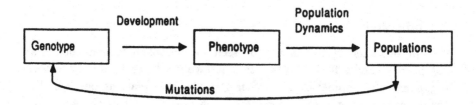

Population Genetics

Population genetics developed during the first part of this century.
It is based on some approximation concerning the phenotype/genotype
relation. The easiest approximation is of course that of Mendeleian
genetics: One gene determines one phenotypic character. We do know
several examples of such one-to-one correspondence and Mendel's first success with peas was based upon the fact that the character smooth or
shrivelled only depends upon a unique gene. In such cases the prediction of the respective alleles proportions inside a population can be easily
calculated using combinatorics.[1]

One to one genotype/phenotype correspondence is the exception rather
than the rule. Genes interact via the products they code for. The fact
that one gene is expressed, which means that the protein it codes for is
actually synthesized, depends upon the expression of other genes. Since the
Operon model[2] proposed by Lwoff, Monod and Jacob, we know explicitly
some interaction mechanisms, by which the product of the expression of
some genes influence in either way — positive induction or repression —
the expression of other genes. One can, at least in principle, imagine a
whole set of interactions among the genes of a given genome, although this
complete set of interactions is still unknown even for the simplest organisms

such as bacterias — the case of the genes of viruses is different since they need other organisms to be expressed.

The central ideal of this paper is, as one may have guessed by its presence in this volume, to represent the complex and unknown set of interactions by a random network of automata and to look for generic features at the level of the phenotype and of population dynamics. The first part of this program concerns cell differentiation and has already been studied by S. Kauffman.[3] We shall adopt the paradigm he proposed in 1969.

Kauffman Model of the Genome

Genes are represented by Boolean automata, which compute their binary state at discrete time intervals as Boolean function of their inputs. The inputs are the binary state of other genes which so influence their expression. If a gene is actually expressed, its state is 1. It is 0 otherwise. Because the details of the real biological interactions are unknown, one chooses a random set of input connections, and the Boolean functions are also chosen randomly for each automaton. Since any Boolean function is defined by its truth table which gives the state of the automaton as a function of its input configurations this choice is simply done by filling the truth table with 0's and 1's randomly sampled. Parallel iteration is chosen which means that all the automata are simultaneously updated. Kauffman interprets the attractors of the dynamics of the network as the different cell types which can exist, although the genome in each cell type is exactly the same. More generally, one can assume that **any phenotypic property** is the result of interactions among the genes and **is a function of the attractors of the network dynamics**. This assertion is the fundamental assumption of the following theory.

Kauffmans's results that are the most important for us can be summarized by saying that two dynamical regimes are observed with such nets, depending upon their input connectivity. The input connectivity is the number of inputs of the automata.

— For automata with low connectivity, say 1 or 2, the periods of the limit cycles of the dynamics are short, and the number of different attractors is also small. These quantities vary as the square root of the total number of automata.

— For connectivities larger than 3, large periods are observed (or guessed when no periods are observed during the time of simulation). They scale as an exponential function of the number of automata.

These two dynamical behaviors remind us from what is observed with continuous dynamical systems: Ordinary attractors such as fixed points or periodic limit cycles on one hand, and 'strange attractors' or chaotic behavior on the other hand.

Robustness

As with continuous dynamical systems one might ask questions about the structural stability of such models. With discrete systems this question translates into what are the changes observed at the level of the attractors when changes are made concerning the initial configuration of a net, or concerning connections and Boolean functions.

Any small change in the net results in some changes concerning the graph of iteration of the net, i.e., the dynamical succession of configurations, as seen in Fig. 1.

The importance of the changes varies according to whether they occur in the transient parts of the graph, or whether they concern the attractors. As seen in Fig. 1b, in change a configuration 29 is followed by configuration 17 instead of 31; both belong to the same basin of attraction and change

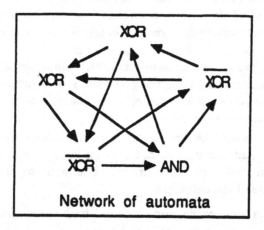

Fig. 1a. A network of five Boolean automata is represented above. The arrows indicate the connections and the function of each automaton is given by its name in logics. Function AND yields 1 iff both inputs are 1, XOR yields 1 iff one input is 1, and $\overline{\text{XOR}}$ yields 0 iff one input is 1.

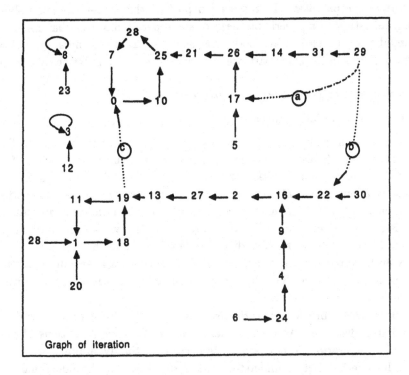

Graph of iteration

Fig. 1b. The network can be in any 32 possible configurations, appearing on the graph of iteration. The digits are the decimal representations of the configurations. The arrows in the graph of iteration figure the time succession of the configurations. If only one bit of a function of the net is changed the iteration graph is partially modified. Three different possible modifications are indicated by the dotted arrows noted a, b and c. They don't have the same importance as explained in the text.

a is of little consequence. In change b, the configuration following 29 is now 22, which belongs to a different attraction basin and the change is more important. But the most catastrophic changes are those like c, which annihilate an attractor. A further graduation would differentiate according to the distances between the destroyed attractor and the new one.

A dynamics is said to be **robust** if the probability of modifying the attractors as a result of small changes concerning the structure of the net is small.

Populations of organisms whose phenotopic properties are robust with respect to changes in their genome due to punctual mutations have specific dynamical behavior such as punctuated equilibria. This motivates our interest in robustness. In the second part of this paper, we shall show that robustness is a generic property of a number of networks that are good candidates for biological modelling such as Boolean nets with low connectivity. In the third part, we shall come back to population dynamics.

Robustness in Networks of Automata

Sensitivity to initial conditions[4,5]

We have already seen that in Boolean nets, low connectivity implies small periods. It also implies low sensitivity to initial configuration. Let us follow the parallel evolution of two configurations which differ at initial time by D automata. D is called the Hamming distance of the two configurations, and $D/N = d$ their distance relative to the number of automata.

For systems with large N and large connectivity, D evolves at large times towards a finite relative distance, however small the initial distance (except for zero distance, of course). d at large times is nearly independent of the initial distance. This behavior is similar to the sensitivity to initial conditions observed in chaotic systems.

In contrast, for low connectivities, the final distance is proportional to the initial distance. As a consequence small initial perturbations of the initial configuration do not propagate because of the dynamics.

This effect is better understood when one looks at the functional organisation of Boolean nets which is better observed on periodic connection structures.

Patterns

The first cellular implementation of Kauffman nets on a cellular lattice is due to Atlan *et al.*[6] It consists in placing Boolean automata with connectivity 2 at the nodes of a square lattice with a connectivity matrix described in Fig. 2. Boundaries are connected together. Since connectivity is 2, these random nets have small periods.

One then observes that during the limit cycles some automata remain stable while others are oscillating (cf. Fig. 3). The set of stable automata is connected and isolates subnets of oscillating automata. The details of the patterns depend upon the initial conditions and are specific to the attractor.

The important point to be observed is that although the network is completely connected by its structure, since all its nodes can be reached from any node by following connecting arrows, when it reaches the limit cycle its

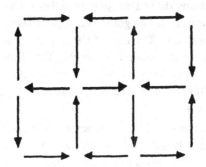

Fig. 2. Pattern of connectivity for 2 inputs cellular automata on a square lattice.

```
. 0 0 0 . . * . 0 1 * . 1 0 1 *          . 1 1 1 . . * . 0 1 * . 1 0 1 *
* 0 1 0 1 0 . . 0 1 * . 1 1 1 1          * 0 1 0 1 0 . . 1 1 * . 1 0 1 0
0 . . 0 0 0 0 . . 0 1 1 * 0 1 1          0 . . 1 1 0 1 . . 0 1 1 * 0 1 1
. . 0 0 1 0 0 * * 1 1 0 * . * .          . . 0 0 1 0 0 * * 0 1 1 * . * .
* * 0 1 * . 0 0 . 0 * * . * * .          * * 1 0 * . 0 0 . 0 * * . * * .
. 0 0 0 . . 0 1 . * 1 0 * . . *          . 0 0 0 . . 0 0 . * 1 1 * . . *
. . * 0 . * 0 0 1 . 1 1 1 . . .          . . * 0 . * 0 0 1 . 1 1 1 . . .
. 1 0 0 . 1 0 . * 0 0 * 1 . . .          . 0 0 1 . 0 0 . * 1 0 * * . . .
0 1 1 1 0 1 1 0 0 * * * 1 0 * 1          . 1 0 1 1 1 1 0 0 * * * * * * .
1 * 0 . 0 1 1 0 1 1 . 1 1 1 0 0          * * . . 0 1 1 1 1 0 . * . * * .
0 0 1 * 1 0 0 1 * . 1 0 . 1 1 1          . * * * 1 0 0 1 * . 1 1 . . * .
1 1 * * 0 1 0 . * * 0 0 * 1 1 0          . * * * 0 1 0 . * * 0 1 * * . .
* . * * 0 0 0 * . . . 0 0 1 0 .          * . * * 0 0 0 * . . . 1 0 1 1 .
* * * . * . . . . * . * . 1 0 . *        * * * . * . . . . * . * . 1 0 . *
. * . . . . * . * * * . 1 0 1 . .        . * . . . . * . * * * . 0 1 1 . .
* . . * * * * * * * . . 1 1 . . .        * . . * * * * * * * . . 1 1 . . .
```

Fig. 3. Pattern of activity of a 16*16 random Boolean net with cellular connections. The same net is observed after it has reached the limit cycle, starting from two different initial configurations. The 0's and the 1's correspond to oscillating automata, while the '.' and the '*' indicate nodes which are invariant during the limit cycle.

full connectivity falls apart. The network is now functionally divided into independent oscillating subunits separated by the stable connected region. This implies that perturbations concerning one node, for instance, a simple flip, do not propagate outside a single oscillating subunit. If they occur in the stable region they are soon damped. We have also tested, for instance, that large perturbations do not propagate outside the subunits: One can 'cut' a part of a network by a line going through its stable region and replace this part by a different net. The rest of the net is not much perturbed and carries on oscillating on the same limit cycle. This functional organisation is also observed in networks with random connections. We have chosen to show it with cellular nets in order to visualize the interactions by the neighborhood relationship.

One can now understand that when an initial configuration is modified in a number of randomly selected automata, one modifies a proportional number of subunits. This proportionality relation is still observed when one measures the distance between the configurations at large times.

In short, random Boolean nets with low connectivity become disconnected into independent subnets when they reach their attractor. As a result, modifying the function of an automaton can only affect the subnet to which it belongs. This is the reason for the smooth variations of the dynamical properties of the nets when one goes across the set of all Boolean nets with a given number of automata.

Other Examples of Robustness

Robustness is also a property encountered in some neural nets and cellular automata.

Neural nets[7] are built according to Hebb's prescription to ensure that predefined configurations, also called references, are the attractors of the dynamics. The same reasons that ensure stability of the reference also ensure that configurations close to the reference in terms of Hamming distance converge towards the reference. Figure 4 shows the robustness of the retrieval as a function of the Hamming distance of the initial configuration to the reference.

To study the robustness with respect to changes in the net, one changes the synaptic intensities which define the net. Punctual mutations among such nets involve only one synaptic condition, and at most two automata. Since changing two automata states increases Hamming distances by two, convergence toward an eventually slightly modified attractor occurs with high probability, which once more guarantes the smoothness of the dynamics. Neural nets have also been shown to be insensitive to a limited "Dilution" (annihilation of connections).[9]

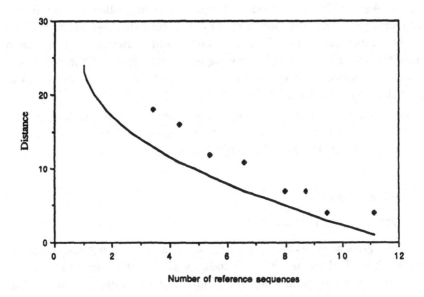

Fig. 4. This figure taken from Weisbuch and Fogelman[8] shows the maximum Hamming distance to the reference which ensures at least a 50 percent chance to return to the reference as a function of the number of reference patterns for a network with 50 automata. Rhombuses are simulation results and the continuous curve is a theoretical result.

In cellular automata, all automata have the same function, and the connection structure is a periodic lattice. Let us consider the case of Boolean automata on a linear lattice with a connectivity limited to the two neighbors and the automaton itself as studied by S. Wolfram.[10] Those Boolean functions with small periods, the class 1 and 2 functions as defined by Wolfram, exhibit robustness with respect to initial conditions. These functions are called forcing since their output might be determined by only one of their inputs. For instance function AND always yields 0 output if one of its inputs is 0, whatever the state of its other inputs. If the left input is 0, for instance, the node is 'deaf' to anything coming from the right. Its right and left sides are then independent from each other. For this reason, these networks are also structured into independent subnets, and are robust with respect to initial conditions as well as to local changes of the Boolean functions of the nodes. (In fact a careful analysis shows that the structuration of random Boolean nets is also due to forcing functions.)

The robustness of some cellular automata with respect to initial conditions has been used to model evolution in a paper by M. Kerzberg and Z. Agur.[11] In their model the genome is modelled by the initial configurations, and the phenotype by the configuration after a finite number of iterations on a cellular chain of Boolean automata. Their fitness character is a function of the Hamming distance to some reference configuration. Population dynamics of such nets exhibit the same behavior as the one I shall describe further.

In conclusion, robustness of dynamical properties at the level of the attractor is a property shared by a wide range of networks of automata, typically those used to model biological organization.

Evolution of Species

The Paradox of Punctuated Equilibria

Although Darwin's theory of natural selection driving Evolution of Species diversified by mutations is today well admitted among scientists, the details of its dynamics are not completely understood. One of the most controversial issues recently discussed is the problem of "**punctuated equilibria**".[12] Evolution can be monitored at two possible levels: A macroscopic phenotypic level, or at a microscopic genetic level. The genome of a living species determines in principle its phenotype, i.e., the set of all its physical properties observable at the macroscopic level. The phenotype refers, for instance, to the size of an animal, to its body confirmation, to its living habits, etc., and what is most important with respect to evolution, to a set of characteristics defining its **fitness**. Fitness defines the ability of an animal to survive, to develop, and to reproduce itself.

The apparent paradox is the following. When evolution is monitored at the gene level, by comparing the difference in homologous proteins between different species, a constant rate of changes of the amino acids in a series is measured. On the other hand, when one follows the changes at the phenotypic level, by measuring physical quantities of fossils in phylogenetic series, discontinuities appear: In other words, evolution is characterized by a succession of very long **stasis** during which the quantity being monitored does not change, separated by short intervals of discontinuous changes with no fossils detected. Most mathematical models based on quantitative genetics could not predict these discontinuities because they were based on a one gene-one character correspondence. Paleontologists proposed two types of explanations to solve the paradox:

— "Catastrophic" events, either climatic, or geologic, or meteorites, etc., which could have suddenly changed the conditions on Earth.

— That mutations responsible for important evolutionary steps were different from those occurring inside a single species. It was proposed that the mechanisms responsible for macro-evolution were different from those responsible for micro-evolution.

We intend to show that no such hypotheses are necessary to explain punctuated equilibria. They only result from the global character of the relation between genotype and phenotype. We shall use Kauffman nets to model the genome, and its attractors to determine the phenotypic characters, such as fitness, which are important in population dynamics.

Equations for Population Dynamics

Let us consider the set of all possible organisms, in which genome i is represented by a network of automata, with a given number of automata N and a given set of connections. Parallel iteration is chosen. The nets differ only in the Boolean functions that are present at each node. The set of all boolean nets with N 2-input automata with a chosen structure of connections, contains 16^N nets. Our purpose is to describe the evolution of populations in this huge space of possible genomes, under several reasonable assumptions concerning death, reproduction, food assimilation and mutations among the organisms. A simple differential system that describes population dynamics among this set of organisms is[13]:

$$\dot{P}_i = a_i P_i - b \sum_j P_j P_i + \sum_v m(t) P_v. \tag{1}$$

The variation in time of the population of each organism i is due to 3 terms:

— A growth term, which is a balance between reproduction and death, with a fitness coefficient a_i characteristic of the organism;

— A constraint concerning resources allocation among the set of all competing organisms (expressed in differential form in the second term);

— A stochastic mutation term, which allows mutations among organisms whose genome differs by only one gene (punctual mutations). The time average of $m(t)$ is m, which is taken as small compared to the fitness terms a_i. $m(t)$ fluctuates on short time scales, but its average on longer time scales does not change with time.

This is of course an oversimplified model: No sex is involved, only punctual mutations are allowed, the only interaction among organisms consists in sharing a common resource, etc.

Competition

If one were to start with all species present with small populations at the origin of time, the dynamics of this system would be quite simple. Let us for a moment neglect mutation terms which are supposed to be small anyhow.

$$\frac{\dot{P_i}}{P_i} = a_i - b \sum_j P_j = a_i - bP_t . \tag{2}$$

The populations of organisms with fitness larger than the second term, bP_t, grow. Because of the increase of P_t, populations that were successful at a given time start becoming unsuccessful because their fitness is now less than bP_t. Only those with larger fitnesses continue growing. Finally the population of the organisms with the largest fitness becomes much larger than any other. The only other surviving organisms are those with a neighboring genome, because of mutations.

Evolution

In fact, evolution of species corresponds to initial conditions with only a few organisms present, none of them corresponding to maximum fitness. Because of mutations, other organisms appear, whose success (i.e., population) depends upon their fitness, upon available resources, and upon history which might favor the appearance of neighboring genomes. In terms of dynamics, we are interested in transients toward stable or metastable states as much as in the equilibrium positions.

As we shall see further on, the mathematical problem of solving the differential system is tractable. But beforehand, the issue of the correspondence between fitnesses and genomes is the big problem. Ideally, we would like to have some model of the ontogenetic development of the organism that would enable us to relate fitness to genome. But embryology is still very far from allowing to build a specific model, and we have to rest on some assumptions. Our basic **conjecture** is the following: The qualitative properties of the dynamics described by the differential system are the same for a large class of distribution of the fitness coefficients, namely those obtained from the dynamical properties of the Boolean nets modelling the genome.

From this conjecture, because of the robust character of these properties, we can assume that fitness varies smoothly in the genome space, which is sufficient to solve the differential system, at least for small values of m.

Solution of the Differential System by Perturbation Methods

The Growth Phase

Let us suppose that at time $t = 0$ only one organism is present. Its population starts increasing, and so do the populations of the neighboring genomes. Those with higher fitnesses increase faster, and their population overcomes those of the original organism. For small values of m, a rather narrow cloud of populations starts drifting towards a close local maximum of fitness. These dynamics are rather rapid.

If m is smaller than b, mutations from a predecessor organism i to a neighboring fitter mutant occur after the population of the predecessor attains saturation. The time interval t between mutations is given by

$$t = \frac{\ln \frac{m}{b}}{a_i} + \frac{b}{ma_i} . \tag{3}$$

For intermediate values of m, such that useful mutations occur prior to saturation, the time interval between successful mutations is

$$t = \frac{\ln \frac{m}{a_i}}{a_i} . \tag{4}$$

In both cases the evolution towards the local maximum of fitness is accelerated: Time between mutations decreases following the increase in fitness.

The Stasis

When a local maximum is reached, a quasi-equilibrium is established. Because of mutations, the genomes close to the maximum have a non-zero population. Following a perturbation method, one computes the successive equilibrium populations starting from the local maximum and going one mutation step away at each time. By keeping in each expression the terms of lower order in m, one obtains the folowing series of equations:

$$\dot{P}_m = a_m P_m - b P_m P_m = 0 \tag{5a}$$

$$P_m = \frac{a_m}{b} \tag{5b}$$

where index m refers to the local maximum.

For the first neigbors with indices v we get:

$$P_v = \frac{mP_m}{a_m - a_v} \qquad (6)$$

which is generalized to distance d, the number of point mutations between the considered genome and genome i:

$$P_i = P_m m^d \sum_c \prod_j \frac{1}{a_m - a_j} \qquad (7)$$

where the sum is extended to all mutation paths going from m to i across different j's.

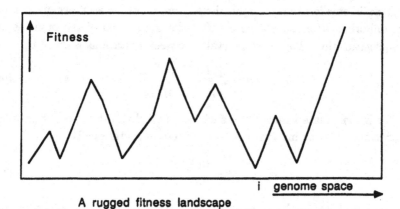

A rugged fitness landscape

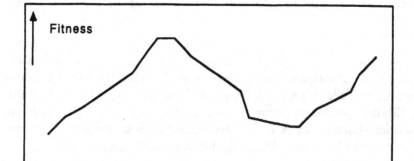

A smooth fitness landscape

For smooth fitness landscapes and small mutation rates, the width of the population distribution is smaller than the distance between local maxima. The valleys of the fitness landscape are deserts, the probability of mutant birth is small and they soon disappear because of their low fitness. The population "cloud" constitutes a **species**, isolated from other organisms, and centered on the local maximum. The notion of species can thus be introduced on the fitness landscape basis, independently from considerations about reproduction gap among different mating organisms.

The lack of population in the valleys is the reason for the existence of stasis in evolution.

From One Peak to the Next Higher Peak

The time to wait until the appearance of a mutant fitter than those already present around the metastable peak is of the order of:

$$t = \frac{1}{mP_m} \prod_i \frac{a_m - a_i}{m} \tag{8}$$

where the product is taken along the steepest line of the lowest saddle point which joins the ancient maximum to the new one. After the birth of the mutant, the cloud of population very soon leaves the ancient maximum and establishes itself around the new one.

Numerical Simulations

We have simulated, for instance, the evolution of populations of organisms with only 6 genes represented by Boolean automata. The graph of interactions is represented below.

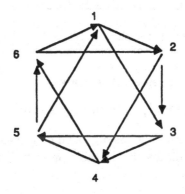

We have selected only functions NAND, OR, AND and NOR among the sixteen Boolean functions. They are further coded respectively 0, 1, 2 and 3. The total number of possible genomes is then 4096. From the graph of iteration of all these nets we have measured global dynamical quantities such as the largest attractor period or the number of different attractors per net. We considered as authorized punctual mutations, i.e,, the mutations that change only one function per genome according to one of the following arrows:

$$\text{AND} <\!\!-\!\!-\!\!> \text{NOR} <\!\!-\!\!-\!\!> \text{NAND} <\!\!-\!\!-\!\!> \text{OR} <\!\!-\!\!-\!\!> \text{AND};$$

in other words, the mutations that change only two bits of the output function. We checked the smoothness of the variation of the two above dynamical quantities when the space of genomes is described by such mutation steps. As an example, the figures for T_i, the largest periods, are given across plane xy3200 in the six-dimensional space. This notation means that automation number 1 has function x, whose codes varies according to the column of the table; number 2 is y; number 3 is 3, which stands for NOR; number 4 is 2 (AND); and the two last automata have function NAND.

x / y	2	3	0	1	2
2	7	1	4	1	7
3	10	1	3	(4)	(10)
0	8	4	(11)	(9)	(8)
1	1	1	1	1	1

Table of the largest period T_i in the plane xy3200 of the hyperspace of genomes. This table shows that even a small network of six automata has a smooth T_j landscape.

Computer simulations represented in Fig. 5 were done with

$$m = 2 \cdot 10^{-5}, \quad b = 10^{-7}, \quad a_i = 0.05^* T_i.$$

We started at time t with only one organism of genome $i = 233200$ and of largest period 10, as indicated in the table present. It is a local maximum of fitness with $a_i = 0.5$.

Four different dynamical regimes can be observed in Fig. 5: The first phase is a rapid expansion of this first organism, and of its two neighbors, of largest period 4 and 8. It lasts some 100 time steps. A first 'species' has appeared (fast evolution).

Once the local fitness maximum is reached, the cloud of population remains centered around it, for 85 998 time steps. A mutant of period 9 occasionally appears (Stasis).

At this time a fitter mutant appears, with code 003200 and period 11. Its population grows exponentially and overcome in a few hundred time steps the previous adapted 'species' which becomes a fossil.

After equilibrium has been reached, a new statis is observed, and so on.

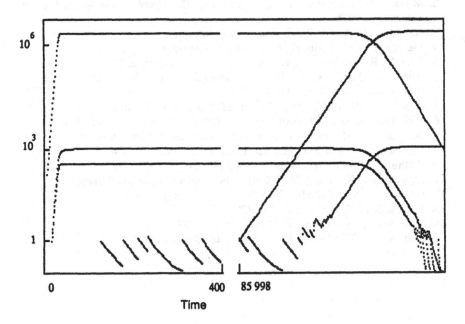

Fig. 5. Evolution of populations simulated on computer. For times smaller than 400, the different curves correspond to genomes of periods 10, 8, 4 and 9, as indicated in the table, by decreasing order of populations. At time 85 998 an organism of period 11 appears and overcomes the others since it has a higher fitness.

Similar results are obtained for a wide range of parameters, and for different dynamical properties determining the fitnesses. The peaks then occur for different genomes, but the semi-quantitative properties of the population dynamics, and the alternation of fast changes and stasis, are the same. We suspect that if we had based our model on neural nets dynamics, and chosen the fitness in similar fashion to the death function used by Anderson and Stein[15] (see also the article by D. Rokhsar in this volume), punctuated equilibria would also have been observed.

This model is thus very general, and it explains quite well the existence of punctuated equilibria. In particular, it is a minimal hypothesis model, in the sense that no extra hypothesis is necessary to explain the succession of stasis and fast changes.

References

1. Ewens, W. J., *Mathematical Population Genetics* (Springer, 1979).
2. Miller, J. H. and Reznikoff, W. S. (editors), *The Operon*, Cold Spring Harbor Laboratory, 1978.
3. Kauffman, S. A. (1969), *J. Theor. Biol.* **22**, 437.
4. Derrida, B. and Weisbuch, G. (1986), *J. Physique* **47**, 1297.
5. Derrida, B. and Stauffer, D. (1986), *Europhysics Lett.* **2**, 739.
6. Altan, H., Fogelman-Soulié, F., Salomon, J., and Weisbuch, G. (1982), *Cybernetics and Systems* **12**, 103.
7. Hopfield, J. J. (1982), *Proc. Nat. Acad. Sci., USA* **79**, 2554.
8. Weisbuch, G. and Fogelman Soulié, F. (1985), *J. Phys. Lett.* **46**, 623.
9. Sompolinsky, H., in *Heidelberg Colloquium on Glassy Dynamics*, ed. van Hemmen, J. L. and Morgenstern, I. (Springer, 1987), p. 485.
10. Wolfram, S., *Rev. Mod. Phys.* **55**, 601 (1983).
11. Agur, Z. and Kerzberg, M. (1986), submitted to *American Naturalist*.
12. Gould, S. J. and Eldredge, N. (1977), *Paleobiology* **3**, 177.
13. Wright, S. (1982), *Evolution* **36**, 427.
14. Weisbuch, G., (1984), *C. R. Acad. Sci.* **298**, 375.
15. Anderson, P. W. (1983), *Proc. Nat. Acad. Sci.* **80**, 3386.

Mathematical Models of Evolution on Rugged Landscapes

Alan S. Perelson

Theoretical Division,
Los Alamos National Laboratory,
Los Alamos, NM 87545, USA

and

Catherine A. Macken*

Department of Mathematics,
Stanford University,
Stanford, CA 94305, USA

Problems involving multi-peaked fitness or energy landscapes have attracted attention in evolutionary biology, structural biology (protein folding), condensed matter physics (glasses and spin glasses) and computer science (combinatorial optimization). Motivated by an important problem in the biology of the immune system, involving the somatic evolution of antibody molecules, we develop and analyze a mathematical model for molecular evolution on rugged landscapes. First, we characterize evolution on random landscapes and use the mathematical results to interpret immunological experiments on affinity maturation by somatic mutation. This analysis provides an explanation for the previously unexplained phenomenon in which mutations initially lead to large improvements in antibody fitness, but later lead to no or little improvement. Second, we show how to extend our mathematical results to a family of correlated landscapes, which we believe will be more appropriate for the quantitative interpretation of immunological data.

1. Introduction

Rugged or multipeaked landscapes are a common underlying feature of many complex systems. They are studied from various viewpoints in the physics of glasses and spin glasses, in the biophysics of macromolecules, in

*Permanent address: Department of Mathematics and Statistics, University of Auckland, New Zealand.

computer science and neural network approaches to combinatorial optimization problems, and in evolutionary biology. Here we outline a mathematical approach to studying the evolution of proteins and nucleic acids, and apply our results to the study of the somatic evolution of antibody molecules. This example is an interesting one because the mutation process is sufficiently rapid that large evolutionary changes can be observed within single animals over a period of a few weeks. Further, modern techniques in molecular biology make it relatively easy to collect large amounts of sequence data with which one can track the evolutionary process. Most importantly, the evolutionary process is fundamental to understanding the dynamics of the immune response and thus characterizing it is of great interest in immunology. However, irrespective of the interest in immunology, the mathematical methods that we employ and the results that we obtain should be of interest in the large variety of fields in which rugged landscapes occur.

To place the particular problem that we will consider in perspective one needs to know a little immunology. When an *antigen* (foreign molecule) is injected into an animal, the animal generally responds by making *antibodies* that can bind to the antigen and lead to its elimination. The effectiveness of the antibody thus depends on its equilibrium binding constant or *affinity* for the antigen. During the course of the immune response the average antibody affinity generally increases with time. Although this phenomenon, called the *maturation of the immune response*, was discovered in the early 1960's,[9] its molecular basis is only now being unraveled. During an animal's first encounter with antigen, the antibodies that are made are encoded by the genes the animal has inherited from its parents. These genes are called germline genes since they are carried by the animals[2] germ cells (sperm and eggs). However, some days into the response, it is found that the B lymphocytes that secrete antibody have mutated their antibody coding genes.[4,21] These mutations are called somatic mutations since they occur in the soma or non-germ cells of the body. The somatic mutations are almost always point mutations that change a single base pair in the DNA that encodes the antibody to another base pair. The mutation rate is unusually high, of order 10^{-3} per base pair per generation.[29,5] As mutations accumulate, the affinity of the antibody increases. After increases typically of one or two orders of magnitude, affinity improvement seems to stop. Further mutations that are seen in animals tend to be silent, affecting the DNA but not the antibody. (Because of degeneracy in the genetic code, about 25%

of base changes lead to no change in amino acid.) For mutations to be detected, cells carrying the mutation must be isolated from the animal. It is believed that mutations that lead to low affinity antibodies do not allow B cells to grow, whereas those that lead to improved affinity lead to increased cell growth. Thus selection can strongly influence the mutations that are seen in an animal, and has led to the general conclusion that the affinity increases in a stepwise fashion with mutation.[20]

We visualize the process of affinity maturation via somatic mutation as a walk on a rugged landscape. Since deleterious mutations are rarely seen, we shall assume that all steps in the walk are uphill. Because a rugged landscape is characterized by having many local optima, a strictly uphill walk will quickly become trapped at a local optimum. We believe that this trapping is the reason that the mutation process stops generating improved mutants after a one to two order of magnitude increase in affinity. The theory that we develop below will predict various statistical properties of walks on landscapes, such as the number of steps required to reach an optimum. Interestingly, many of the predicted features of such walks are in qualitative or quantitative agreement with observations made about affinity maturation.

2. A Model of Affinity Maturation

The primary structure of proteins and nucleic acids can be specified by a sequence of N symbols, chosen from an alphabet of a letters, with each letter representing an amino acid or a nucleotide. Sequence space, S_N, is the set of all a^N possible sequences of length N. A point mutation changes one letter in the sequence to another. A sequence of point mutations can then be viewed as inducing a walk in sequence space. To each sequence one can assign a fitness. In the case of antibodies, the fitness has a natural interpretation as the affinity of the antibody for the immunizing antigen. John Maynard-Smith[28] suggested modeling protein evolution by a sequence of mutational steps each of increasing fitness, downhill steps being neglected due to strong evolutionary selection. In the case of antibody evolution this suggestion seems quite appropriate as Kocks and Rajewsky[20] report stepwise increases in affinity with mutation. Also, rare mutations that increase affinity are shared among independently derived antibodies indicating the strong influence of selection.[10,36,39,12] The study of evolution in sequence space begun by Maynard-Smith has been continued and refined by Jacques Ninio,[30] Manfred Eigen,[8] John Gillespie,[13] Zvia Agur and Michael Kerszberg,[1] Stuart Kauffman,[15–17,19] Edward Weinberger,[37,38] Peter Schuster,[11,34,35] Luca Peliti,[2,7] and their coworkers. With the exception of recent work by Weinberger, most approaches have relied on computer simulation rather than mathematical analysis. Our approach focuses on the development of models that can be studied analytically.

Somatic mutations are restricted to the variable (V) region of the antibody molecule. This region can be described at either the protein or nucleic acid level. If antibody V regions are viewed at the protein level, then $a = 20$ and N is approximately 230, the number of amino acids in the combined heavy and light chain V regions. If we view antibody V regions at the level of DNA, then $a = 4$ and N is approximately 700. Our theory is independent of the mode of describing antibodies. Two sequences that differ in one position only are called *one-mutant neighbors*. Two-mutant and j-mutant neighbors can be defined analogously. Since a single point mutation changes one letter in a sequence, evolution by point mutation generates a connected walk among one-mutant neighbors in sequence space. The number of one-mutant neighbors of a sequence is defined to be D. In protein design experiments, $D = 19N$, since each possible amino acid substitution creates a unique one-mutant neighbor. However, for proteins, such as antibodies, evolving by point mutations in DNA, we replace the actual number of one-mutant neighbors at the DNA level, $3N$, by a smaller effective number of one-mutant neighbors in order to accommodate restrictions in the genetic code. To calculate D, we assume 25% of nucleotide substitutions are silent, and that antibody V_L and V_H regions jointly contain approximately 700 nucleotides, i.e., $D = .75 \times 700 \times 3 = 1575$. We use $D = 1500$ as a typical value.

To each sequence in sequence space we assign a fitness, i.e., its affinity for the immunizing antigen. Although the affinity of each antibody sequence could be determined experimentally, this is clearly not feasible given the large number of possible sequences. One would like a means of predicting affinity from sequence, but a structurally based calculation is not yet available. Since affinity cannot be predicted from sequence with certainty, we simply assign each antibody a fitness randomly selected from a specified probability distribution. We do not believe that fitnesses are random functions of sequence. However, this choice of relationship describes one extreme possibility. When fitnesses are randomly assigned, neighboring sequences can have relatively very different fitnesses, and the landscape will be *rugged*. Even if the fitness function is not random the landscape can be rugged. Thus a study of random fitness functions provides qualitative information about the class of rugged landscapes. A similar philosophy has proven valuable in condensed matter physics, where the study of systems with random energy functions has led to insights into the properties of spin glasses (cf. Derrida[6]). Once a fitness is randomly assigned to each sequence, a landscape is generated, with the fitness being the height of the landscape (Fig. 1). The fitness distribution may be normal, lognormal, uniform, etc. If one chooses a distribution that is tightly peaked around its mean, then one is assuming that a single point mutation will most likely not change affinity very much in absolute terms. Conversely, using a uniform distribution implies that a single mutation has equal probability of generating any allowed affinity. In our model, fitness is assigned to sequences at random from a probability distribution $G(u)$. Interestingly, most of our results are independent of the distribution G.

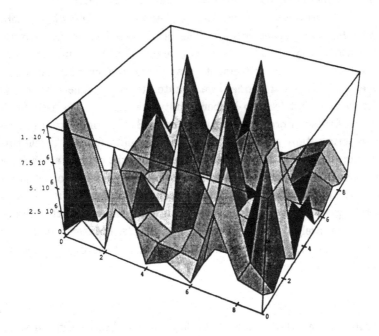

Fig. 1. A schematic illustration of a rugged fitness landscape. The fitness values were chosen from the lognormal distribution, Eq. (1), with $\mu = 6$ and $\sigma = 2/3$.

The nature of a walk on a landscape is as follows. At any stage on a walk, single-mutant variants of the current antibody are tested in random order until the first neighbor having a higher fitness is attained. The walk then moves to this new point in sequence space and the testing process starts anew. Thus, each move on a walk changes the current antibody sequence to a neighboring one. As there are D one-mutant neighbors of each point in sequence space, at most D different single mutant variants can be tested. If no fitter variant is found among these different one-mutant neighbors, the process stops, as the walk has reached a local optimum. Thus, from a single starting sequence our model tracks the lineage of a single antibody.

Besides the application to immunology, which we discuss in detail below, our model may be viewed statistically as a sequential sampling scheme. We sample at random from a distribution G. As each sample is drawn, it is compared with the current maximum sample value. If it is larger, then the maximum is updated; if it is smaller, another sample is taken. If D samples have been taken since the most recent update, then the process stops. This sampling scheme has been studied by Jirina[14] and Saunders.[31-33]

There are two events in the walking process to be carefully distinguished. One is a *step* to a higher fitness. The other is a *trial* of a one-mutant neighbor which may or may not have a higher fitness. This latter event might not lead to any movement away from the current position in sequence space. In our model, we equate steps to higher fitness with point mutations that lead to higher affinity antibodies. We assume that expression of these improved mutations causes B-cell clonal expansion that can be experimentally observed during an immune response. Trials which result in antibodies with lower fitness than that of the current antibody are assumed not to be expressed or to lead to cell lineages that do not expand significantly. Why expansion does not occur is not known, but the empirical facts appear to be that mutants with lower fitness than the germline antibodies are rarely if ever observed. For example, Manser[25] analyzed 20 clonally related mutants, 19 of which had roughly equal or higher affinity than the unmutated precursor.

3. Results

We have proven that many of the key results about walks on rugged landscapes are independent of the particular distribution of fitness values.[24] Whenever this distribution affects the model's predictions, we assume that affinities have a lognormal distribution, truncated to lie within feasible bounds ($10^2 - 10^{10}$ M^{-1}). This assumption is equivalent to assuming that the free energy of binding has a normal distribution (see Ref. 23 for further justification of this assumption).

We denote antibody affinities by the random variable U. We denote the probability distribution $\Pr(U \leq u)$ by $G(u)$, and assume that a corresponding density, $g(u)$, exists. In particular, for our choice of a lognormal distribution

$$\frac{dG(u)}{du} = g(u) = \frac{K}{u} \exp\left\{-\frac{1}{2}\left(\frac{\log_{10} u - \mu}{\sigma}\right)^2\right\}, \qquad (1)$$

where K is a constant chosen to ensure that $g(x)$ integrates to 1 between 10^2 to 10^{10}.

Characteristics of the Landscape

Consider a sequence of length N, composed of elements selected from an alphabet of size a, with the number of one-mutant neighbors $D = (a-1)N$. Let S_N denote the number of local optima. Then the expected number of optima[15] $E(S_N) = a^N/(D+1)$ and the variance[22] $\text{var}(S_N) = a^N[D - (a-1)]/[2(D+1)^2]$. For large N, $\text{var}(S_N)/E(S_N) \cong \frac{1}{2}$, which reflects some non-randomness in the distribution of local optima (i.e., local optima cannot be one-mutant neighbors of one another). Further, for any fitness distribution $G(\cdot)$, as $N \to \infty$, the number of optima is normally distributed with the above mean and variance.[3] Thus, there are a large number of local optima. Because local optima are frequent (on average, one out of every $D + 1$ sequences is an optimum), walks toward optima tend to be short, and only a small portion of sequence space is searched.

Characteristics of Walks to a Local Optimum

A walk progresses by mutating a sequence of a given fitness at single sites until the first fitter mutant is obtained or until no new mutations are possible, whichever occurs first. If a sequence is the starting point of the walk, then each of its D one-mutant neighbors might lead to a higher fitness, whereas if a sequence was produced by a single mutation of some less fit sequence, then only $(D-1)$ one-mutant neighbors might possibly lead to a higher fitness. We denote the starting fitness of a sequence u_0 and later fitnesses u. The results given below are exactly true only for infinite-sized sequence spaces since we ignore the possibility of a walk intersecting itself. However, this is not a practical limitation for the biological problems of interest since $a = 4$ or 20 and $N > 100$ guarantee that the size of the sequence space is extremely large. The variable $\theta = D(1 - G(u))$ is useful to describe a boundary layer near the global optimum $G(u) = 1$. When $\theta \simeq O(1)$, the mathematical behavior of the model changes character. Terms such as $G^D(u)$ are transcendentally small for $\theta \gg 1$.

Fitness after k steps toward a local optimum. Let $U_k(u_0)$ be the fitness after k steps without reaching a local optimum in the preceding $k-1$ steps, starting from fitness u_0. We define the probability distribution $F_k(u; u_0) = \Pr(U_k(u_0) \le u)$, where $\frac{d}{du} F_k(u; u_0) \equiv f_k(u; u_0)$. Since we start at fitness u_0, with probability $1 - G^D(u_0)$ we can take one step, and the attained fitness $u > u_0$ has probability density function $g(u)$. Suppose $k \ge 2$. For the mutation process to achieve a fitness u in k steps, two independent events must have occurred: First, the process must have achieved a fitness u', $u_0 < u' < u$, in $(k-1)$ steps, with density function $f_{k-1}(u')$; second, the process must have tested i less fit one-mutant neighbors before producing a one-mutant neighbor with fitness $u > u'$, with density function $G^i(u')g(u)$, $0 \le i \le D - 2$, where $G^i(u') \equiv [G(u')]^i$. Hence, writing G_0 for $G(u_0)$,

$$f_1(u; u_0) = g(u) \sum_{i=0}^{D-1} G_0^i = g(u)\frac{1 - G_0^D}{1 - G_0}, \quad u > u_0,$$

and

$$f_k(u; u_0) = g(u) \int_{u_0}^{u} f_{k-1}(u'; u_0) \frac{1 - G^{D-1}(u')}{1 - G(u')} du', \quad u > u_0, \ k \ge 2. \quad (2)$$

The recursion Eq. (2) can be solved to obtain[24]

$$f_k(u; u_0) = \frac{1 - G_0^D}{1 - G_0} \frac{g(u)}{(k - 1)!} \left(V(G) - V(G_0)\right)^{k-1}, \quad k \ge 1, \quad (3a)$$

where

$$V(G) = \int_0^G \frac{1 - q^{D-1}}{1 - q} dq. \quad (3b)$$

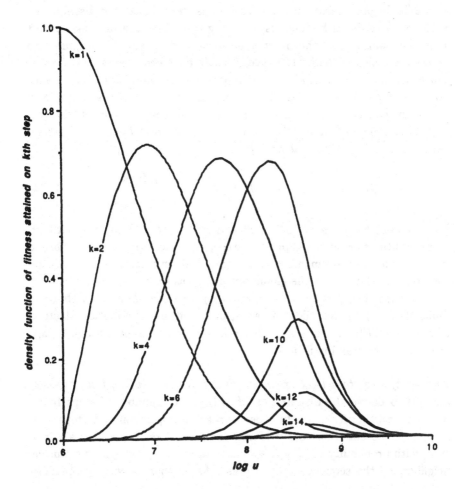

Fig. 2. The probability that fitness u is attained on the kth step of a walk of at least k steps versus $\log_{10} u$. The figure is based on the assumption that walks start at a sequence with affinity $u_0=10^6$, that $D = 1500$, and that affinities are distributed according to a lognormal distribution, Eq. (1), with $\mu = 6$ and $\sigma = 0.8$.

The density function $f_k(u; u_0)$ depends on the distribution of fitnesses as well as the starting fitness u_0. In Fig. 2 we display the density function for various values of k for $D = 1500$ and u chosen from the lognormal distribution (1) with $\mu = 6$ and $\sigma = 0.8$. The starting affinity $u_0 = 10^6$, a value close to affinity of the unmatted antibody in the experiments of Manser.[25,26] In a previous paper of ours[23] a similar graph is given with

$\sigma = 2/3$. In both cases the peak, as well as the mean of the distribution, shifts to the right as k increases, implying that there is a progressive shift to higher affinities as the number of steps k increases. The figure also shows that the area under the fitness distribution decreases as k increases. For a particular value of k the area represents the probability that a path continues for at least k steps. As k increases the chance of reaching a local optimum on or before the kth step increases, and hence the curves "shrink".

The nth moment, $n = 1, 2, \ldots$, of $U_k(u_0)$, conditional upon the kth step being on a path of length at least k is

$$E[U_k^n(u_0)| \text{ path has } \geq k \text{ steps}] = \frac{\int_0^1 u^n f_k(u; u_0)du}{\int_0^1 f_k(u; u_0)du}. \qquad (4)$$

By evaluating Eq. (4) for $G(u_0) = 0.5$ and $g(u)$ lognormal, we find that high values of fitness are attained in a few steps, with succeeding steps leading to only marginal improvements.[23] Starting at different fitness values outside the boundary layer near the global optimum has little long-term effect.[23]

Although these results depend on $G(u)$, one can easily show that the moments for the percentiles, or ranks, of the fitness distribution attained after k steps, $E[F_k^n(u; u_0)]$ are independent of $G(u)$ except through $G(u_0)$, the rank of the starting fitness.

Number of steps to a local optimum. One can use $f_k(u; u_0)$ from Eqs. (2) and (3) to derive the distribution of $W(u_0)$, the number of steps to an optimum starting from u_0. In order to reach an optimum in k steps, two independent events must occur: First, the process reaches a fitness u in k steps with probability $f_k(u; u_0)$; second, all remaining $(D - 1)$ one-mutant neighbors of the sequence have fitnesses lower than u, with probability $G^{D-1}(u)$. Thus

$$p_W(k; u_0) \equiv \Pr[W(u_0) = k] = \int_{u_0}^1 f_k(u; u_0)G^{D-1}(u)du, \quad k \geq 1. \qquad (5)$$

Hence

$$p_W(k; u_0) = \frac{1 - G_0^D}{1 - G_0} \frac{1}{(k-1)!} \int_{G_0}^1 [V(x) - V(G_0)]^{k-1}x^{D-1}dx, \quad k \geq 1. \qquad (6)$$

For any starting fitness u_0, the distribution of the lengths of paths to an optimum is dependent on the underlying distribution of the fitnesses *only*

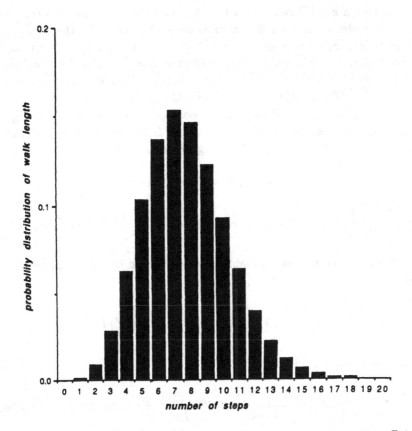

Fig. 3. The probability, $p_W(k; u_0)$, that a walk to a local optimum, starting at affinity 10^6, ends on the kth step, for $D = 1500$.

through $G(u_0)$, an observation also predicted by the theory of dependency graphs.[3] Figure 3 exhibits the probabilities $p_W(k; u_0)$ for a walk starting at $u_0 = 0$ with $D = 1500$.

We have computed $p_W(k; u_0)$ for various values of k and D. The distributions have long upper tails (cf. Fig. 3), which suggests that on rare occasions a walk may be rather long. In conjunction with the conclusions about the fitness attained after k steps (see Eqs. (3) and (4)), we realize that long walks result from an initial rapid increase in fitness followed by a "muddling around" at high fitnesses. Thus the optimization problem clearly obeys the law of diminishing returns.

Although Eq. (6) can be used to compute the moments of $W(u_0)$, we take an easier route via a first step analysis. Let $W(u)$ be the number of steps to an optimum from $u \geq u_0$. Consider the events that can occur immediately after the process has reached u. For example, if $u > u_0$, then with probability $G^{D-1}(u)$ this fitness will be an optimum; or with probability $\sum_{i=0}^{D-2} G^i(u)g(u')du'$, the process will take another step of increasing fitness to $u' > u$, with $W(u')$ the random number of steps remaining to reach an optimum. Thus, using the notation $D^*(u)$ where $D^*(u_0) = D$ and $D^*(u) = D - 1$ for $u > u_0$, and writing D^* for $D^*(u)$,

$$
W(u) = \begin{cases} 0 & \text{with probability } G^{D^*}(u), \\[2ex] 1 + W(u') & \text{with probability} \dfrac{1 - G^{D^*}(u)}{1 - G(u)} g(u')du', \quad u' > u. \end{cases}
$$

(7)

Invoking the law of total probability, we obtain an implicit expression for the expectation of $W(u)$, $\mu_W(u)$, namely

$$
\mu_W(u) = \frac{1 - G^{D-1}(u)}{1 - G(u)} \left\{ 1 - G(u) + \int_u^1 \mu_W(u')g(u')du' \right\}, \quad u_0 < u. \quad (8)
$$

The solution to Eq. (8) is

$$
\mu_W(u) = \frac{1 - G^{D-1}}{1 - G} e^{-V(G)} \int_G^1 e^{V(y)} dy, \quad u_0 < u. \quad (9a)
$$

An analogous derivation applies to the calculation of $\mu_W(u_0)$, the mean number of steps to an optimum from starting fitness u_0. For this case, up to D one-mutant neighbors might need to be tried on the first step of the walk, instead of $D - 1$. Thus[24]

$$
\mu_W(u_0) = \frac{1 - G_0^D}{1 - G_0} e^{-V(G_0)} \int_{G_0}^1 e^{V(y)} dy. \quad (9b)
$$

The standard deviation of $W(u_0)$, $\sigma_W(u_0)$, can be calculated similarly.[24] For $D(1 - G_0) \gg 1$, $\mu_W(u_0)$ and $\sigma_W^2(u_0)$ can be shown to approach $\ln(D-1) + \ln(1 - G_0) + 1.1$ and $\ln(D-1) + \ln(1 - G_0) + 0.26$, respectively. Notice these results are independent of the fitness distribution G, apart from the constant $G(u_0)$.

Using these results, we predict that the mean number of steps to an optimum is rather small, generally under ten, even for proteins of substantial size. The size of $\sigma_W(u_0)$ indicates that substantial variability is to be expected in experimental measurements of the number of mutational steps until an optimal fitness is attained. In the case of antibodies, with $D = 1500$, we predict that the number of steps to an optimum starting with an antibody which has an affinity of 10^6 (i.e., the median of the affinity distribution) is 7.7 ± 5.2 (mean $\pm$ two s.d.). This prediction is rather interesting immunologically since the number of replacement mutations seen in somatically mutated antibodies is of order 10, with variations running between 2 and 20. Manser[27] estimates the average number of mutations at six. Thus, this admittedly crude theory does capture the major features of the observed data.

Total number of trials to reach a local optimum. A step is taken on a walk only when it leads to a higher fitness. As many as $D - 1$ neighbors may have to be examined before a step is taken. We denote each attempt to take a step, a trial. Let $t(u)$ be the total number of distinct trials and $T(u)$ the total number of not necessarily distinct trials on a walk from fitness u to a local optimum. The total number of trials can be related to evolutionary time if the rate at which variants are generated is determined. As walks attain higher fitness, on average more trials occur before a higher fitness neighbor is found. At an optimum, D^* distinct trials are made but none lead to a higher fitness. Via first-step analysis we find for starting fitnesses outside the boundary layer, i.e., for $D(1 - G(u)) \gg 1$,[24]

$$E[t(u)] \sim 0.781D, \tag{10a}$$

$$E[t^2(u)] \sim D^2, \tag{10b}$$

and

$$E[T(u)] \sim \kappa D, \tag{11}$$

where

$$\kappa = \int_0^\infty [1 - e^{-E_1(z)}](e^z - 1)z^{-1}dz = 1.224\ldots,$$

and E_1 is the exponential integral. As with the number of steps to a local optimum, these results are distribution-free.

Table 1. Statistics on the number of trials needed to leave fitness u, where $G(u) \times 100$ is the percentile of u. [From[24] Eq. (3.19).]

	number of trials	
$G(u)$	mean	standard deviation
0.5	2.0	1.4
0.6	2.5	1.9
0.7	3.3	2.8
0.8	5.0	4.5
0.9	10.0	9.5
0.95	20.0	19.5
0.99	100.0	99.5

The results are intriguing. First, for $D = 1500$, on average over 1200 variants need to be tested in order that an antibody attains its optimum. Most of the trials are involved in moving from a high fitness to an optimum.[24] Thus during the initial phase of an immune response, when affinities are low, only small numbers of trials are required to take a step. Later in the response, larger numbers of trials are required (Table 1). If the number of antibody-secreting B cells expressing a particular antibody V gene is less than the number of trials needed to find a fitter variant, our model suggests that antibodies will not reach a local optimum by the end of an immune response but will only have evolved to a high fitness. Thus the observation that somatic mutation generally leads to one or two orders of magnitude increase in affinity (say 10^5 to 10^6 or 10^7 M^{-1}) but not to very high affinities (e.g., 10^9 M^{-1}), may be explained either by the attainment of a low fitness optimum or by the response terminating before a sufficient number of variants are tested. Second, included in the trials taken to reach an optimum are those leading to higher fitness. For walks beginning at the median fitness $G(u_0) = 0.5$ and with $D = 1500$, we expect that about 0.7% of distinct trials lead to a higher fitness. Thus over the entire walk less than 1% of mutations should be improvements. Although there are as yet no direct tests of this prediction, this number is quite plausible biologically. Third, the total number of trials needed to attain an optimum is insensitive to the starting fitness u_0 as long as u_0 is outside the boundary layer. This last result is particularly surprising and must in part reflect the fact that most of the trials required to find an optimum occur near the optimum where almost all one-mutant neighbors have lower fitness.

Comparing $E[T(u)]$ with $E[t(u)]$, we see that the evolutionary process is quite efficient. For walks starting in the outer region, on average only 36% of all attempted mutations during the walk will be previously tried sequences. Most of these repeated mutations occur when the process is very close to a local optimum.[24]

Distribution of values of attained local optima. In order that a fitness u be a local optimum reached by a path starting at u_0, we require that a path reach u without becoming trapped at a lower-valued local optimum, an event having probability $f_k(u; u_0)$ for $k = 0, 1, \ldots$ [cf. Eqs. (2) and (3)]; and that u be a local optimum, an event having probability $G^{D-1}(u)$. Hence, the density function for attained local optima, $f_{\text{opt}}(u)$, is

$$f_{\text{opt}}(u) = G_0^D \delta(u - u_0) + g(u) \frac{1 - G_0^D}{1 - G_0} G^{D-1}(u) e^{V(G(u)) - V(G_0)}, \quad u \geq u_0.$$
(12)

Provided $g(1) > 0$ and $D(1 - G_0) \gg 1$, $f_{\text{opt}}(u)$ has the asymptotic form $f_{\text{opt}}(u) \sim g(u) \frac{D}{\theta} e^{-\theta - E_1(\theta)}$, where $\theta = D(1 - G(u))$ and E_1 is the exponential integral.[24] Thus, most of the optima are clustered in a boundary layer near $G(u) = 1$. Even though the process gets trapped at a local optimum, the evolutionary process is still very efficient. The fraction of sequences with fitness higher than the obtained optimum is on average

$$\int_0^1 (1 - G(u)) \frac{D}{\theta} e^{-\theta - E_1(\theta)} g(u) du \approx 0.62/D.$$

4. Correlated Landscapes

The work summarized above has presupposed that fitnesses are assigned at random to sequences. This clearly cannot be correct and thus it is important to study molecular evolution on correlated landscapes. Kauffman and coworkers have suggested a model for a family of correlated landscapes.[16,17] The model, called the NK model, is similar to a k-ary spin glass model and has been studied extensively by numerical simulation.[18,19] Here we suggest an even simpler model, the *block model*, which can be studied analytically. The block model, as the NK model, has a tunable degree of correlation among the fitnesses of neighboring sequences. In this model, we view each sequence in sequence space as composed of a set of B functionally independent blocks, each of possibly different length n_i, $i = 1, \ldots, B$, where $\sum_i n_i = N$, the length of the entire sequence. We assume each block makes a random fitness contribution U_i to the total fitness U of the sequence. One can consider two possible schemes for combining fitnesses: a multiplicative scheme, $U = \prod_i U_i$; and an additive scheme, $U = \sum_i U_i$. In antigen-antibody interactions, assuming that blocks are functionally independent implies that the total binding free energy is the sum of the contributions made by each block. Thus if binding energy were used as a fitness, an additive scheme would be appropriate. Conversely, if affinity is used as the fitness of a block, then the product scheme seems more appropriate since affinities multiply when the free energy contributions of each block are added.

Changing the number of blocks changes the characteristics of the fitness landscape. If $B = 1$ our model reduces to the random landscape model that we studied above. If $B = N$, so that the number of blocks equals the number of letters in the sequence, the landscape is totally correlated.[16] Intermediate levels of correlation occur for $1 < B < N$.

To give the flavor of the type of results one can get from this model, consider the computation of the mean number of local optima as a function of B. Now, $d_i = (a - 1)n_i$ is the number of one-mutant neighbors of block i. A sequence is a local optimum if its fitness is higher than that of its one-mutant neighbors. At a local optimum each block must be at a local optimum, otherwise a mutation could increase the fitness of the entire sequence. Let S_{n_i} denote the number of local optima of a block of size n_i. Then, $E(S_{n_i}) = a_i^n/(d_i + 1)$. The expected number of local optima of the entire sequence $E(S_N) = \prod_i^B E(S_{n_i})$. If all B blocks are of size $n = N/B$, then each has $d = (a - 1)n = D/B$ neighbors, and $E(S_N) \simeq a^N/(D/B)^B$. Comparing with the result for one block, we see that the number of local optima *decreases* by the factor B^B/D^{B-1} due to the block structure, suggesting a commensurate increase in the lengths of walks to a local optimum. Changing the distribution of fitnesses within a block changes the degree of correlation among neighbors.

Because of the independence of the blocks we can easily study the statistics of the lengths of walks to an optimum, the number of mutations it takes to reach an optimum, etc. Preliminary work suggests that a two block model will lead to improvements in understanding the features of antibody evolution. A full account of the block model will be published elsewhere.

Acknowledgements

The asymptotic results quoted in this chapter were derived in collaboration with Patrick Hagan. This work was performed under the auspices of the U.S. Department of Energy and partially supported by the LDRD program at Los Alamas National Laboratory, N.I.H. grant AI28433 to A.S.P. and the Santa Fe Institute through their Theoretical Immunology program.

References

1. Z. Agur and M. Kerszberg, "The emergence of phenotypic novelties through progressive genetic change", *Am. Nat.* **129** (1987) 862–875.
2. C. Armitrano, L. Peliti and M. Saber, "A spin-glass model of evolution", in *Molecular Evolution on Rugged Landscapes: Proteins, RNA and the Immune System*, edited by A. S. Perelson and S. A. Kauffman (Addison-Wesley, 1991), pp. 27–38.
3. P. Baldi and Y. Rinott, "Asymptotic normality of some graph related statistics", *J. Appl. Prob.* **26** (1989) 171–175.
4. C. Berek and C. Milstein, "Mutation drift and repertoire shift in the maturation of the immune response", *Immunol. Rev.* **96** (1987) 23–41.
5. S. H. Clarke, K. Hüppi, D. Ruczinsky, L. Staudt, W. Gerhard, and M. Weigert, "Inter- and intraclonal diversity in the antibody response to influenza hemaglutinin", *J. Exp. Med.* **161** (1985) 687–704.
6. B. Derrida, "Random-energy model: An exactly solvable model of disordered systems", *Phys. Rev.* **B24** (1981) 2613–2626.
7. B. Derrida and L. Peliti, "Evolution in a flat fitness landscape", *Bull. Math. Biol.* **53** (1991) 355–382.
8. M. Eigen, "Macromolecular evolution: dynamical ordering in sequence space", in *Emerging Syntheses in Science, Proceedings of the Founding Workshops of the Santa Fe Institute*, edited by D. Pines (Addison-Wesley, 1988), pp. 21–42.
9. H. Eisen, "Affinity maturation: a retrospective view", in *Molecular Evolution on Rugged Landscapes: Proteins, RNA and the Immune System*, edited by A. S. Perelson and S. A. Kauffman (Addison-Wesley, 1991), pp. 75–82.
10. S. Fish, E. Zenowitch, M. Fleming, and T. Manser, "Molecular analysis of original antigenic sin. I. Clonal selection, somatic mutation, and isotype switching during a memory B cell response", *J. Exp. Med.* **170** (1989) 1191–1209.
11. W. Fontana, W. Schnabl and P. Schuster, "Physical aspects of evolutionary optimization and adaptation", *Phys. Rev.* **A40** (1989) 3301–3321.
12. M. Fleming, S. Fish, J. Sharon, and T. Manser, "Changes in epitope structure directly affect the clonal selection of B cells expressing germline and somatically mutated forms of an antibody variable region", submitted for publication.

13. J. H. Gillespie, "Molecular evolution over the mutational landscape", *Evolution* **38** (1984) 1116–1129.

14. M. Jirina, "Sequential estimation of distribution-free tolerance limits", *Selected Translations in Mathematical Statistics and Probability* **1**, (Amer. Math. Soc., 1961), pp. 145–156.

15. S. A. Kauffman and S. Levin, "Towards a general theory of adaptive walks on rugged landscapes", *J. Theor. Biol.* **128** (1987) 11–45.

16. S. A. Kauffman, E. D. Weinberger and A. S. Perelson, "Maturation of the immune response via adaptive walks on affinity landscapes", in *Theoretical Immunology, Part One*, edited by A. S. Perelson (Addison-Wesley, 1988), pp. 349–382.

17. S. A. Kauffman and E. D. Weinberger, "The NK model of rugged fitness landscapes and its application to maturation of the immune response", *J. Theor. Biol.* **141** (1989) 211–245.

18. S. A. Kauffman, "Adaptation on rugged fitness landscapes", in *Lectures in the Sciences of Complexity*, ed. D. L. Stein (Addison-Wesley, 1989), pp. 527–618.

19. S. A. Kauffman, *Origins of Order: Self Organization and Selection in Evolution* (Oxford Univ. Press, 1992).

20. C. Kocks and K. Rajewsky, "Stepwise intraclonal maturation of antibody affinity through somatic hypermutation", *Proc. Natl. Acad. Sci. USA* **85** (1988) 8206–8210.

21. C. Kocks and K. Rajewsky, "Stable expression and somatic hypermutation of antibody V regions in B-cell developmental pathways", *Ann. Rev. Immunol.* **7** (1989) 537–559.

22. C. Macken and A. S. Perelson, "Protein evolution on rugged landscapes", *Proc. Natl. Acad. Sci. USA* **86** (1989) 6191–6195.

23. C. Macken and A. S. Perelson, "Affinity maturation on rugged landscapes", in *Molecular Evolution on Rugged Landscapes: Proteins, RNA and the Immune System*, edited by A. S. Perelson and Stuart A. Kauffman (Addison-Wesley, 1991), pp. 93–118.

24. C. Macken, P. Hagan and A. S. Perelson, "Evolutionary walks on rugged landscapes", *SIAM J. Appl. Math.* (1991), in press.

25. T. Manser, "Evolution of antibody structure during the immune response. The differentiative potential of a single B lymphocyte", *J. Exp. Med.* **170** (1989) 1211–1230.

26. T. Manser, "Maturation of the humoral immune response: A neo-Darwinian process?", in *Molecular Evolution on Rugged Landscapes: Proteins, RNA and the Immune System*, edited by A. S. Perelson and S. A. Kauffman (Addison-Wesley, 1991), pp. 119–134.

27. T. Manser, "The efficiency of antibody affinity maturation: can the rate of B-cell division be limiting?", *Immunol. Today* **11** (1990) 305–308.

28. J. Maynard-Smith, "Natural selection and the concept of a protein space", *Nature* **225**, (1970) 563–564.

29. D. McKean, K. Huppi, M. Bell, L. Standt, W. Gerhard and M. Weigert, "Generation of antibody diversity in the immune response of BALB/c mice to influenza hemagglutinin", *Proc. Natl. Acad. Sci. USA* **81** (1984) 3180–3184.

30. J. Ninio, *Molecular Approaches to Evolution* (Princeton Univ. Press, 1983).

31. S. C. Saunders, "Sequential tolerance regions", *Ann. Math. Stat.* **31** (1960) 198–216.

32. S. C. Saunders, "On the sample size and coverage for the Jirina sequential procedure", *Ann. Math. Stat.* **34** (1963) 847–856.

33. S. C. Saunders, "Some applications of the Jirina sequential procedure to observations with trend", *Ann. Math. Stat.* **34** (1963) 857–865.

34. P. Schuster, "Potential functions and molecular evolution", in *Chemical to Biological Organization*, edited by M. Markus, S. C. Mueller and G. Nicolis (Springer-Verlag, 1988), pp. 149–165.

35. P. Schuster, "Optimization and complexity in molecular biology and physics", in *Optimal Structures in Heterogeneous Reaction Systems*, edited by P. J. Plath (Springer-Verlag, 1989), pp. 101–122.

36. J. Sharon, M. L. Gefter, L. J. Wyoski, and M. N. Margolies, "Recurrent somatic mutations in mouse antibodies to *p*-azophenylarsonate increase affinity for hapten", *J. Immunol.* **142** (1989) 596–601.

37. E. D. Weinberger, "A rigorous derivation of some properties of uncorrelated fitness landscapes", *J. Theore. Biol.* **134** (1988) 125–129.

38. E. D. Weinberger, "Correlated and uncorrelated fitness landscapes and how to tell the difference", *Biol. Cybernetics* **63** (1990) 325–335.

39. L. J. Wysocki, M. L. Gefter and M. N. Margolies, "Parallel evolution of antibody variable regions by somatic processes: Consecutive shared somatic alterations in V_H genes expressed by independently generated hybridomas apparently acquired by point mutation and selection rather than by gene conversion", *J. Exp. Med.* **172** (1990) 315–323.

The Spin-Glass Analogy in Protein Dynamics

Robert H. Austin and Christine M. Chen

Department of Physics,
Princeton University,
Princeton, NJ 08544, USA

1. Foreword: Just What Problem Are We Trying to Solve?

It is the conceit of this paper that the spin-glass model provides a useful way of looking at the dynamics of a complex, heterogenous macromolecule such as a globular protein. Many of the concepts discussed in protein dynamics are related to issues that arise in spin-glass physics. A spin glass is a *solid* object which undergoes some sort of freezing transition at some temperature. Likewise, a globular protein is a *solid* object which also undergoes some sort of a glass transition at some temperature, as we will show. Further, a spin glass at low temperatures reveals a complex *distribution* of states on a rough potential surface. The modification of the energy landscape by the random and frustrated interactions of the magnetic spins modifies the thermodynamics of the intermediate states, and the relaxation of the dense set of states reveals a highly non-exponential magnetic relaxation below the glass temperature. These kinds of physical effects are also seen in proteins, as we will show.

We also think that, aside from the physics questions, the material discussed here is of biological significance. In any biological reaction there are (at least) two physical quantities of biological interest. One quantity, which I think we all agree is important, is the thermodynamics of the reaction. What are the free energies of the reactants, intermediate substates, and the products? That class of proteins called enzymes, of most interest in this paper, are true catalysts and thus do not change the free energies of the initial and final states but rather change the *rate* at which the process occurs. This brings us to the other quantity of interest in biological reactions: how does the protein change the *rate* of the reaction? In fact,

179

the two aspects are closely coupled together: the protein changes its conformation as it binds to a substrate, the energy barrier is lowered in this bound state complex and the product dissociates. Implicit in this scenario is the importance of the conformational flexibility of the protein. In many cases the protein must move to allow entry of the substrate, and then during a reaction must move between distinctly different conformations in order to achieve non-linear coupling between reactant and product.[1]

Thus, proteins need to move between different conformations in many if not most situations in order to work. We believe that the dynamics of a spin glass *above* the glass transition temperature are directly related to the dynamics of the conformations of the protein above its own glass transition, and that by understanding the subtle magnetization fluctuations above the glass transition we will understand something about the dynamics of a protein molecule.

Over the years RHA has really pissed off some very eminent physicists by giving talks on the spin glass analogy to protein structure and dynamics. Being too naive to be discouraged, we have spent a long time mulling over the wounds caused by the slings and arrows of outraged theorists. To be honest with the reader, at least RHA was pretty skeptical that the spin-glass model was of any relevance when he began to write this paper two years ago. However, it now seems to us from the evidence that we have collected here that actually the analogy may be quite strong. We still think that there is something to the analogy and we will try to prove it here. We hope that in drawing our perhaps strained analogy between spin glasses and proteins that some sort of over-riding principles emerge to help us understand the conformational energetics and dynamics of these complex biological macromolecules. We hope that you don't get *too* pissed off, but if we are doing our job you will be a little pissed.

2. What is a Spin Glass?

Since the thermodynamics and dynamics of a spin glass are quite different from a structural glass, with important physical consequences, it is important right here at the start to get things defined crisply. An example of a real, physical spin glass is a solid metallic alloy which contains a dilute concentration of paramagnetic impurities in a host conductor. An example would be iron doped into gold.

A spin glass is very different from the structural glass that one normally thinks of when the word "glass" is mentioned.[2] A structural glass is a *liquid* above its glass transition temperature and undergoes a kinetic arrest to an amorphous structural state before it can reach its global energetic minimum crystal configuration.[3] As we mentioned above, a spin-glass is a solid for all temperatures above and below the glass transition, but which undergoes a static transition to a disordered spin configuration due to frustrated spin-spin interactions.

Now, how does a spin glass form? Consider an array of spins in space interacting with each other via some coupling J_{ij} between the ith and the jth spin, and in the presence of an external magnetic field H. The Hamiltonian for such a system is:

$$\mathcal{H} = -1/2 \sum_{ij} J_{ij} S_i S_j - H \sum_i S_i^2 . \tag{1}$$

In a metallic spin-glass, the spin in this dilute solution of magnetic spins interacts with neighboring spins via the RKKY (Ruderman-Kittel-Kasuya-Yosida) potential[4-6] which arises from the interaction of the magnetic spin with the conduction electrons of the host metal. In the RKKY interaction the magnetic spin interaction with the conduction electrons of the metal gives rise to a spin-spin coupling parameter J_{ij} between the ith and jth spin which falls off algebraically with distance and is effectively *random in sign* due to the random distribution of positions of the spins in the host metal.

We have to make a careful distinction here, following Binder and Young,[8] between spatial and bond disorder. In spatial disorder there is a randomness to the separation between the sites. Thus, even though the RKKY potential is smoothly analytic, there will be an effective random interaction between the sites. A bond-disorder model could have the sites on a perfectly regular lattice, but allows the interaction J_{ij} between the sites to be random in sign. This case, while somewhat unrealistic compared to a "real" spin glass, is usually easier to calculate. It is not clear to these authors if the properties of these two models of disorder are significantly different.

In any event, the randomness of the sign of the coupling gives rise to the very important concept of *frustration*.[10] Frustration is extremely important in understanding why a spin-glass forms a disordered state at low temperatures, and if our analogy is to work, must play a crucial role in the protein problem as well. A clear discussion of frustration, as well as a very readable discussion of many of the problems in spin-glass physics can be found in the recent article by Fisher *et al.*[9]

Frustration arises when there are competing interactions of opposite sign at a site.[7] Frustration is easy to demonstrate as we show in Fig. 1: for 3 spins on a triangle with the given couplings it is impossible for the third spin to satisfy its neighbors' requests for up/down orientation. A truly frustrated lattice cannot have its disorder transformed away; it is inherently complex.[8] This criterion has ruled out other models of spin glasses which were really "ferromagnets in disguise",[8] and exhibited some but not all of the characteristics of spin glasses.

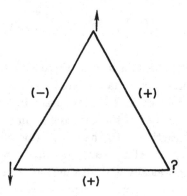

Fig. 1. A simple triangular lattice with 3 spins. The coupling interaction J_{ij} between the spins is shown as + (parallel) or − (anti-parallel). The spin in the lower right-hand corner is frustrated in that it cannot simultaneously satisfy both coupling constants.

A frustrated system has no global ground energy state, but rather a large number of nearly isoenergetic states separated by large energy barriers. In other words: imagine that you take a large interacting system and split it up into two parts. Minimize the energy of the two separate parts. Now bring the two parts back together. A frustrated system will not be at the global energy minimum. There are a number of physical consequences that arise from frustration. The primary consequence of this glass transition that we wish to stress here is *the presence of a dense multitude of nearly isoenergetic states separated by a distribution of activation energy barriers.* This is due to the presence of frustrated loops which cannot be easily broken by any simple symmetry transformation. This complexity inherently leads to distributions of relaxation times.

There are some limiting cases to the complexity which allow one to make calculations. In the limit of a large, purely random system the statistics of the states should become Gaussian. Edwards and Anderson proposed a simple distribution for couplings $P(J_{ij})$ in a spin glass[12] is:

$$P(J_{ij}) = (2\pi J_b)^{1/2} \exp(-(J_{ij} - J_o)^2/2J_b^2). \qquad (2)$$

In the Edwards-Anderson model this interaction is assumed to be finite ranged, while in the Sherrington-Kirkpatrick model[11] the range is unphysically chosen to be infinite. Thus, the Edwards-Anderson spin glass model seems to be more physically realistic than other models, and is also the model that has been used by Stein in modeling proteins.

Many of the characteristics of a spin glass can be found in similar systems which share the phenomena of local disorder and *competing* interactions among neighbors, and this is where the power of the spin-glass analogy is revealed. The example of greatest relevance to proteins may be the insulating *orientational* glass, which occurs in dielectric solids which are doped with dipole impurities such as CN^-.[13] Here the magnetic spin impurities are replaced by electric dipole moment impurities if there are ferro/antiferro competing couplings, or electric quadrupole moment impurities if the strain field of the lattice is the coupling interaction. Now, there must be frustration in such a system if the spin glass model is to apply. If we simply had a sample consisting of electric dipoles then there would be nothing but ferroelectric coupling and no frustration.

It is well known that no molecule can have a *static* electric dipole moment. In the case of the CN^- group, the electric dipole moment of the CN arises from the interaction of the CN group with the strain field of the lattice which breaks symmetry. This strain field is spatially anisotropic and can exhibit the same sort of oscillatory sign dependence as does the RKKY interaction in the case of a spin glass in a metallic host. In fact the analogy is amusing: in the spin glass the magnetic dipoles interact with the conduction electrons which then interact with other spins, while in the elastic dipole case the dipoles interact with the phonons of the lattice which interact with the adjacent dipoles. A clear description of the elastic dipole glass can be found in the article by Grannan *et al.*[14] Just for the record and to make it look like we know what we are talking about let us give an example of an elastic Hamiltonian. Define the dipole moment Q_{ij} between two oppositely charged centers of charge Q_o and index i and j as:

$$Q_{ij} = Q_o[r_i r_j - \frac{1}{3}\delta_{ij}] \tag{3}$$

then the long range strain field Hamiltonian is:

$$H = -\frac{1}{2} \sum_{r_1 \neq r_2} Q_{ij}(r_1) J_{ijkl}(r_1 - r_2) Q_{kl}(r_2) \tag{4}$$

where $J_{ijkl}(r_1 - r_2)$ is the anisotropic coupling of the dipoles, which plays the role of the random interaction J_{ij} in spin glasses. See Grannan *et al.* for details and further references. There are more exotic examples of frustrated insulating glasses. An example is the so-called proton glasses, which are formed by mixed crystals of hydrogen-bonded ferroelectric and anti-ferroelectric materials,[15] which also can show frustration.

It is one step removed to claim that orientational glasses found in dielectric solids resemble spin glasses, which are magnetic systems. It is yet another step removed to claim that the complex "solid" formed by a condensed amino acid polymer resembles an orientational glass, yet that is what we are going to claim in this article.

3. What a Protein Glass Might Be

Now, let's see if we can contrast a spin glass with a globular protein and find similarities. There are 20 different amino acids commonly found in a globular protein, and they are of highly variable chemical composition. This is to be contrasted with another common biological polymer, DNA, which is chemically much more homogeneous (although structurally it is revealing considerably more variety than had been assumed). Thus, while DNA under physiological conditions is found to be in a B-helix form, proteins under physiological conditions assume a staggering variety of conformational shapes, depending upon the amino acid composition.

Even within one configuration of N amino acid sequences, there are probably exponentially ($e^{\alpha N}$) many ways to fold the protein into a globular configuration. Even in the absence of the elastic interaction we mentioned above perhaps one can say that because of the complex array of amino acids present the interaction J_{ij} between the ith amino acid and the jth amino acid is likely to be random in sign, in analogy to the RKKY potential in spin glasses. We don't know if that is a legitimate thing to say.

Now, of course, the X-ray crystallographers present us with very nice pictures of "the structure" for a given protein. Aha, you say, the present paper is worthless, proteins have a unique structure and don't look like a glass at all! There are three points to make: (1) Jesus, you didn't read carefully what we said above! Spin glasses and orientational glasses occur in quite regular *solid* lattices. (2) We are more concerned with the small structural variations within the broad set of a given framework. These variations can well be hidden within the highly massaged structures that the computers spit out. (3) Crystallization of a protein selects out a subset of the total protein structures in solution, and crystallization forces can drive the system into a common structure. Computer folding simulations indicate that the structures fold into many separate energy minima which are quite distinct from one another. Thus, we do not even at present know if "the structure" seen by X-ray crystallography is a global minimum structure or a selected one. The level to which the reader wishes to draw the spin glass analogy resides on several possible levels.

It is appropriate at this point to raise the question of the size of the protein molecule. As is made very clear in the review article of Fisher,[9] theory in the true spin-glass systems assumes an infinite size sample. In the proteins, the molecules are very definitely of finite size. For example, a small protein such as myoglobin has a molecular weight of approximately 18,000 daltons and a radius of approximately 25 Å. Such a small radius immediately says that many of the approximations used by the theorists will not work.

Does this then mean that there are no phase transitions in proteins? Even small proteins have very well defined denaturation temperatures where over the range of only a degree or so the globular protein changes to a random coil.[16] These changes of state in my opinion are sharp enough to merit the label of a "phase transition". As is always true in the biological physics of macromolecules, we must work by analogy and try to accommodate the physics as best we can. To demand perfect rigor is to abandon the field altogether.

Below this denaturation temperature we believe that the globular protein can be in a very large number of structurally different spatial configurations. At room temperature in solution (that is, fully hydrated) the protein is believed to rapidly jump between the thermally accessible conformation states. We would expect (hope) that below some temperature T_c that the protein molecules no longer can jump between these conformational states and each protein molecule becomes "frozen" in some particular conformation. Note that there is no good reason from what we have said so far to assume that (a) a conformational distribution exists, (b) that the protein can rapidly jump between these conformations or (c) that a glass transition can happen. The evidence for this will come later.

If, however, such a glass transition can occur then the entire conformational space formed by the ensemble of protein molecules with a given amino acid sequence and general structure, which was once rapidly sampled by each protein molecule, is now time invariant in the sense that the protein molecule can no longer sample the states. The space can be considered to be a glass in the spin-glass sense if the two criteria of *randomness* and *frustration* have been satisfied. We hasten to point out that no one to our knowledge has ever clearly demonstrated that frustration occurs in a protein structure. There is only one honest thing to do: we now need to look *experimentally* at what a spin glass looks like and then try to contrast the experimental signature of a spin glass to what is observed in proteins.

4. Properties of Spin Glasses

4.1. Magnetic Susceptibility

A great deal of theory has been written concerning the consequences of frustration and disorder on the magnetization M of a spin glass vs. temperature T. When kT is considerably greater than the mean interaction energy $< J^2 >^{1/2}$ between the spins we expect that the magnetization of the spin glass should obey a simple Curie-Weiss law[17]:

$$M(t) \propto 1/T . \tag{5}$$

This simple temperature dependence is called free-spin paramagnetism and is due of course to a single spin S interacting with the magnetic field B. Imagine, however, that there is spin-spin interaction. If the interaction is of a ferro- or antiferro-magnetic type then below the Curie temperature the magnetization becomes very large due to the net alignment of the spins along some **K** vector. Since all the spins point in the same direction the system can be said to be in one macroscopic state.

However, for a spin glass there is no net alignment of the spins although the spins are no longer free to point in any direction given the statistics of the Boltzmann relation. Since there is no structural change in the system if any thermodynamic phase transition exists we would expect that the transition will not be like a simple first or second order phase transition. Indeed, while above the critical temperature it is observed in spin-glasses that the magnetization does follow a $1/T$ susceptibility below the glass transition it is observed that there is a *cusp* in the magnetization followed by a roughly linear T dependence of the magnetization vs. temperature,[18] as we show in Fig. 2.

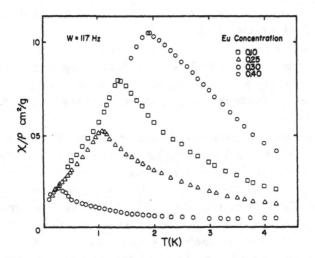

Fig. 2. The real part of the magnetic susceptibility as a function of temperature for $Eu_xSr_{1-x}S$. The frequency of measurement ω was 117 Hz. This figure is taken from the work of Maletta and Felsch.[69]

The presence of a cusp in the susceptibility followed by a *decline* is evidence that the spins are no longer free but instead have been constrained in some direction. However, since there can be no net alignment of the spins in a frustrated system the magnetization falls to zero at low temperatures, unlike a ferromagnetic or paramagnetic system.

As Binder and Young express it, we can characterize a spin glass at low temperatures by two equations. One says that a spin has some orientation in space which when averaged over time t is not zero:

$$\langle \mathbf{S}_i \rangle_t \neq 0 \tag{6}$$

but that there is no net alignment of the entire spin ensemble:

$$\frac{1}{N} \sum_i \langle \mathbf{S}_i \rangle_t \exp(i\mathbf{K} \cdot \mathbf{R}_i) = 0. \tag{7}$$

Presumably, at the critical temperature T_c the spins undergo a glass transition to a time invariant disordered phase. The actual definition of the glass transition T_c is rather murky, since there is no discontinuous change in any thermodynamic properties such as susceptibility or specific heat, and the value at which the cusp in the susceptibility is reached depends on the frequency f at which the measurements are made. In general the higher the frequency the lower is the temperature at which the kinetic arrest seems to occur. However, a plot of the cusp vs. frequency f usually yields a well-defined extrapolated T_c at $f = 0$.[19]

Since the temperature T_c can only be obtained by extrapolation to $f = 0$, it is a matter of controversy as to whether a spin glass undergoes a true thermodynamic phase transition, or instead simply experiences a kinetic runaway where the relaxation times to the true ground state become unreasonably long.

The above reasoning would seem to indicate that the "glass transition" is just a trivial freeze-out of relaxation. However, it was pointed out by Kauzmann quite a while ago[20] that as the glass transition is approached the entropy of the system falls so steeply with temperature that at some temperature T_k the entropy of the liquid is LESS than the entropy of the crystal, which is disturbing. However, the glass transition seems to arrive in the nick of time like the cavalry to keep the entropy of the glass greater than the entropy of the liquid. This view has been modified since. In any case, as Stein has pointed out in a popular article,[21] it is unclear at present whether there really is a thermodynamic glass transition or just a kinetic arrest. Of more general concern is the metastability of glasses in general: if there is no true underlying thermodynamic phase transition, then all glasses can be viewed as metastable systems out of equilibrium.[22] When it comes to "living" molecules, this is a matter of supreme importance.

4.2. Specific Heat Measurements of Spin Glasses

A key aspect of any glass is the existence of a *dense distribution* of states. The specific heat C_p of a solid is a measurement involving a sum over all the states of the system and is sensitive to the distribution of the states, and changes in the structure of the system which change the states. If a system undergoes a structural transition then the specific heat is discontinuous at the transition temperature. Even *structural* glasses, where there is a deeper lying crystalline phase which the supercooled liquid would like to reach, are capable of showing "steps" in the specific heat when the glass is annealed for long periods of time below the glass transition temperature.[2] However, in a spin glass or orientational glass there is no underlying structural state of lower energy hence one would not expect to find discontinuities in the specific heat. Rather, one expects to see effects rather like those seen in the susceptibility measurements discussed earlier, namely a *peak* in the specific heat at some temperature followed by a decline at lower temperatures.

Since we don't expect discontinuities in the specific heat, but rather wide peaks the signature of the glass transition is tricky experimentally. One has to be a bit careful here in looking for broad bumps and proclaiming a spin-glass transition has occurred. Prof. Angell has reminded us that the specific heat of a simple two level system with level splitting Δ actually has a wide maximum at $kT \sim \Delta$, and is of course zero for $T >> \Delta$ and $T << \Delta$[17] (this is known as the Schottky anomaly in doped magnetic systems). Thus, a wide and broad peak in the specific heat is not necessarily indicative of a spin-glass transition. However, in a two level system the specific heat decreases exponentially for low temperatures due to the fixed gap Δ between the two levels. We will see that this is not true for a spin glass due to the distribution of energy levels.

The specific heat of a regular insulating crystalline solid has two major components. In the Debye model the phonons in the solid contribute at low temperature (kT much less than the Debye energy as given by the shortest wavelength mode) give a $C_3 T^3$ dependence to the specific heat. The presence of optical phonons, as well as acoustic phonons, can be taken into account via the Einstein relation for the freeze-out of the highest frequency mode.[17] Empirically, a good fit to specific heat data can be obtained by fitting the specific heat at constant volume to:

$$C_V(T) = C_3 T^3 + C_E[\Theta(E/T)] \tag{8}$$

where C_3 is the Debye contribution and $\Theta(E/T)$ is the Einstein function for a solid with a fundamental energy spacing E in the harmonic approximation.

In the case of a spin glass with a metallic host the situation is more difficult because the specific heat of the conduction electrons at low temperatures is itself linear in T. However, it is possible to remove the electronic contribution to the specific heat and determine the magnetic contribution of the specific heat. When this is done the magnetic contribution to the specific heat for a magnetic impurity with fixed level splitting Δ should decrease as $\exp(-\Delta/kT)$ as we discussed above. However, in a disordered systems such as a spin glass it is found that this exponential fit doesn't work at low temperatures. The specific heat falls much more slowly than an exponential low temperatures, and seems to fit rather well to a simple linear dependence on T, as we show in Fig. 3. The origin of this linear specific heat, one signature of a spin glass, is a matter of active theoretical concern. One early suggestion that still seems to have support is the model of Anderson *et al.*[23,24] that there exists a dense *distribution* of quasi-energetic two level systems. The distribution of tunneling transition rates between the two level systems is believed to give rise to the *linear* specific heat observed in these materials.

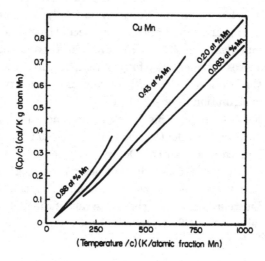

Fig. 3. The magnetic contribution of the specific heat of CuMn vs. temperature. For details, refer to the article by Binder and Young[8] and the original work by Brodale *et al.*[70]

4.3. Distribution of Relaxation Rates and Energy Levels in Spin Glasses

In a spin glass there are local clusters characterized by a local magnetization M_i. If an external magnetic field is applied to this glass, consisting of this cluster distribution, then the longitudinal relaxation of the spins would not be expected to follow a single exponential time decay but instead will have a distribution of relaxation rates due to the distribution of cluster sizes.

There are several ways to see that such a distribution of relaxation times occurs in spin glasses. In frequency space, the real and the imaginary parts of the magnetic susceptibility have been measured. Concentrating on the complex susceptibility for example, we expect that if there is one relaxation time τ then the general form of the complex susceptibility is:

$$\epsilon''(\omega) = \omega\tau \frac{\epsilon_T - \epsilon_S}{1 + \omega^2\tau^2} \tag{9}$$

where ϵ_T and ϵ_S are the isothermal and adiabatic susceptibilities respectively.[25] In reality a large distribution of relaxation times $g(\tau)$ is required to fit the data:

$$\epsilon'' = \int [\epsilon_T - \epsilon_S] \frac{g(\tau)d\ln\tau}{1 + \omega^2\tau^2}. \tag{10}$$

Empirical fits of the distribution show a marked widening of the distribution at the glass phase transition, as we show in Fig. 4.

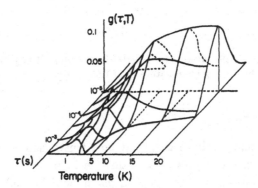

Fig. 4. The distribution of relaxation times $g(\tau, T)$ as a function of time τ and temperature T for the spin glass $(CoO)_{0.4}(Al_2O_3)_{0.1}(SiO_2)_{0.3}$. From the work of Wenger.[71]

In the time domain highly non-exponential effects are seen, as we would expect. The idea here is simple: cool the spin glass to some temperature T and then abruptly apply a small magnetic field H. How fast does the system respond to this perturbation? If there is a single relaxation time then the relaxation process will be exponential in time. Figure 5 shows some typical relaxation time courses as a function of temperature: the process is highly non-exponential, and in fact looks like a power law decay:

$$\sigma_R(t) \propto t^{-n(T,H)} . \tag{11}$$

Other forms of decay laws are possible, and will be discussed in the sections on hierarchical structures and ultrametric spaces.

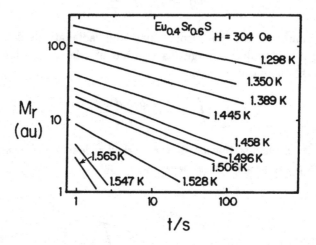

Fig. 5. A log-log plot of the saturated magnetization vs. time of $Eu_{0.4}Sr_{0.6}S$ at various temperatures near the freezing temperature T_f of 1.55 K. From the work of Ferre *et al.*[72]

4.4. Hierarchical States

Up to this point we have discussed how frustration in the spin-spin interactions can give rise to a glass transition below which the spins are no longer able to respond to the applied magnetic field, and we have very briefly discussed how the relaxation times of the system seem to diverge below the glass transition temperature and are given by some sort of a distribution of relaxation times. The most interesting analogy to be drawn to biological systems still awaits us, however. A protein in a glass state is effectively a dead protein. We would like to know the *dynamics* of the relaxation of the system above the glass transition. It is not enough then to state what the distribution of ground state energies are, since the very existence of a glass transition implies that there must be *very large barriers* between the local minima in free energy. In any kinetics problem one wants to know the pathway by which the system can relax to other local minima. We will now show, as one would suspect, that this issue is also quite deep in spin-glass physics and new important concepts emerge.

The first "new physics" we want to very briefly discuss is the concept of a hierarchical space, which is a form of evolutionary tree. You can "prove" that a set of states forms a hierarchical space if a reasonable definition of distance (not Euclidian!) can be found to characterize the path one must take to get between different spin states. If the distance is very large, then "yuh can't get thar from here".

Intuitively, one way to explain hierarchical states is to look at a Rubik's cube. Suppose you have some random the color distribution (state ψ_r), and you'd like to go back to the ordered color state ψ_c. If you could arbitrarily turn any of the colors, going back to the desired state would be trivial and quick. However, because of the construction of the cube there are large free energy barriers between states that are not "close" to the one you are in: you must flow back over the *allowed* states in some very slow process in order to arrive at where you wanted to be. This distribution of allowed states that are close in "distance" and forbidden states separated by a large "distance" is a characteristic of a hierarchical distribution of states.

Now, let's think about a spin glass. Consider a particular spin glass configuration α. The individual spins in the α configuration can be labeled by S_i^α, which in a one-dimensional case can be viewed as a series of $+1$ (up) or -1 (down). Now, consider how you would change the spin state to another state β. Inside of randomly flipping spins, one could compare the two spin configurations and leave alone all the spins that pointed in the same direction, but flip spins that point in opposite directions. For example, suppose that we found that we had to flip the 4th spin of state α. By flipping that spin up or down, one oscillates between two "nearby" states. The bifurcation between these two "nearby" states can be seen as a branch point above the two states. If we restrict ourselves to single spin flips, it is possible to construct an "evolutionary tree" via single spin flips to a common ancestor from which we could descend to the state β. Such a structure that flows from a common ancestor via intermediate states is called a hierarchy. The most common analogy is that tree outside your window (those of you from Illinois or Arizona can refer to a picture in a book). All the branches of a particular tree are related to a common ancestral trunk. The distance between any two branch tips is the physical distance along the tree surface you have to travel to reach the other branch tip.

As we stated, in any hierarchy you must have some quantitative way to characterize the distance between different states in the hierarchy. In a spin-glass, a convenient way to parameterize the similarity between the two different states might be given by the overlap $q^{\alpha\beta}$ between two states:

$$q^{\alpha\beta} = \frac{1}{N} \Sigma_i S_i^\alpha S_i^\beta . \tag{12}$$

Essentially this definition of similarity is a measure of the probability that the two spin configurations match. States that are close to one other will have a value for $q^{\alpha\beta}$ that is close to unity, while states that are far away will have a value near zero. Thus, the "distance" between two states would be given by:

$$d_{\alpha\beta} = \frac{1}{2} N (1 - q^{\alpha\beta}) . \tag{13}$$

Next, a big leap of faith. We might expect the energy barrier between the states would be related in some way to the distance between the states. The actual quantitative relationship must depend on the functional form of the spin-spin interaction. It is reasonable, though unjustified, to assume that they should scale together.

It is believed that the Sherrington-Kirkpatrick (SK) model of a spin glass[11] forms a hierarchical space.[26] Unfortunately, this case may well be pathological because of the infinite range of the spin-spin interaction. In fact, Huse and Fisher[28] have questioned whether a real spin glass could have any of the qualities that the SK model has. The more physical Edwards-Anderson type of spin-glass involves finite-ranged coupling constants between the spins has also been proposed to form a hierarchical system,[27] but the issue is unfortunately unresolved.

The pay-off from all of these musings comes when we start to consider the origin of the observable non-exponential time dependence of things like remanent magnetization in the spin glass below the phase transition temperature. We explained this in the above sections by assuming that a multitude of energy barriers existed, but we did not really justify the existence of the distribution. In the hierarchical scheme, it is possible for a distribution to arise from the many branches of the tree, and by considering diffusion from branch to branch one also has hopes of actually doing a dynamics calculation on this lattice.

For example, Palmer, Stein, Abrahams and Anderson[29] have performed an interesting examination of hierarchically constrained dynamics, in an attempt to understand why some sort of power law was often seen in such systems. In hierarchically constrained dynamics all of the states are assumed to have the same ground state energy, but we wish to observe the diffusion of the probability density $P(t)$ give that we start in one particular state. PSAA were able to show that in various situations kinetics could be observed that fit a variety of non-exponential curves, including power law decays as we discussed in the relaxation rate section, or the Kohlrausch decay law:

$$\sigma_R \propto \exp(-t^\beta) \tag{14}$$

where β is a number less than 1. As we have pointed out, relaxation dynamics in disordered systems often seem to fit such a decay law.

4.5. Ultrametricity

Given that the spin states of at least some spin glasses form a hierarchy, we can finally ask what are the mathematical and physical consequences of this hierarchy. Suppose we ask how one can pass from one spin state to another? Since the spins act in a cross-coupled way, with attendant frustration "clashes" occurring between certain configurations, randomly flipping the spins is likely to lead along high-activation energy paths that are unlikely to occur.

A consistent and logical approach would be to work through the hierarchical tree of states from one state to another. In this way one always goes through states that are closely related to another, and hence presumably travels over minimum energy routes.

If we define the "distance" D between any two states to be the number of generations that one must go back to find a common ancestor, then the concept of distance takes on a decidedly non-Euclidian turn. In fact, in such a space the distance between any three points x,y and z satisfies the inequality:

$$d(x, z) < \max[d(x, y), d(x, z)]. \tag{15}$$

A space which satisfies this relationship is called an ultrametric space.[30] In this simple expression lies a great deal of subtle mathematics. It isn't our purpose in this brief review to go into the complexities of these mathematical ramifications, since in fact (surprise!) the purpose of this paper is to draw the analogy of spin-glass physics to proteins. We will try to address here aspects of ultrametricity that may help us understand the dynamics of protein conformational relaxation and the ordering of protein conformational states.

One striking effect of ultrametricity is the granular nature it gives to different spin configurations. For example, one of the major differences between an ultrametric space and a Euclidian metric is its lack of intermediate states. You can't have three points on a straight line since if points A,B, and C are on a straight line with one meter separation between A and B and one meter between B and C, then we will violate our inequality stated above: A and C can be no more than 1 meter apart in an ultrametric space, rather than the two meters we demand here.

A consequence of the above section is an important theorem that will be important to us in the protein section. It concerns balls, where a ball is defined as all those sets of configurations that are closer than some distance D from each other. The theorem is: any two balls must be either disjoint or contained within the other, that is, no ball can have parts of itself contained in other balls. In the tree analogy, no branch can belong to two separate trunks. This also implies that any two balls of equal radius must be either disjoint or identical. To see this, consider a ball of unity radius. Let point A be at the center of the ball and point B be on the sphere surface. There can be no point outside the ball less than 1 unit from B, because then it would also be at most 1 unit from A, and thus in or on the ball, by our definition of distance. Thus, the sets of spheres are necessarily disjoint. We will return to this important point in the section on proteins.

As we mentioned above, it is not yet clear that the metastable low energy states of a finite-range, three dimensional spin glass forms an ultrametric space. The criterion for ultrametricity is quite formal and mathematical, hence even *thinking* of applying such a concept to a protein must seem like sheer folly, and may well be. However, as we will see, there are several aspects of ultrametricity which can be "tested" in protein dynamics simulations.

Finally, we will briefly discuss the question of diffusion in an ultrametric space. We have previously discussed diffusion in a hierarchical space and found that we expect to get non-exponential kinetics. It should come as no surprise to the reader that the non-exponential kinetics are also expected in the more confined ultrametric space. The paper that most directly addresses this question is by Ogielski and Stein.[31] The ultrametric space that Ogielski and Stein chose to work in was one in which the metric topology was formed by activation energy barriers. That is, they assumed that the bifurcation in the space is formed by a hierarchy of activation energy barriers Δ_i which can be ranked in order of increasing magnitude:

$$\Delta_1 \leq \Delta_2 ... \leq \Delta_k .... \tag{16}$$

In this scheme the ultrametric distance between two sites is related to the total activation energy that must be surmounted in going from one site to another. Several different activation energy rankings in the ultrametric space were studied. The simplest case of equal barriers δ gave the simple result in the limit of an infinite number of sites:

$$P_0(t) \sim -t^{-T \ln 2/\Delta} \tag{17}$$

where $P_0(t)$ is the probability of finding site 0 occupied at time t, assuming that $P_0(0) = 1$. Note that we expect a temperature dependent power law in this case, with the slope of the power law linearly dependent on temperature.

5. Spin Glasses and Elastic Orientational Glasses

We are trying to walk the reader over from the cool and abstract beauties of the spin glass to the wet and wild world of the protein. To ease the reader in this cultural shift we want to discuss here how the dielectric glasses resemble spin glasses in terms of experimental observable. As we mentioned in the early parts of this paper, there is a clear physical analogy between the dipole and quadrupole electric moment systems and the spin glasses, so the correspondences found here should not be too surprising.

5.1. Dielectric Susceptibility

As you will recall, the magnetic susceptibility in spin glasses has a cusp at the transition temperature and then falls below the transition. Importantly, similar behavior is also observed in *orientational glasses*. The dielectric susceptibility is one physical constant which is sensitive to ordering in the lattice of the dipoles and is directly analogous to the magnetic susceptibility. We show in Fig. 6 the expected peak in the dielectric susceptibility in the response of a typical orientational glass. A peak in the real dielectric constant as a function of temperature is observed as the orientational glass is cooled, and the maximum of this peak is observed to be frequency dependent.[32] The real component of the dielectric constant falls sharply as the sample is cooled below the critical temperature, indicating again that the dipole moments assume a *local* orientation but no global one due to the random interaction and frustration, which kills any ferroelectric transition. It seems clear from this figure that at least the orientational glasses have similar behavior to the spin glasses.

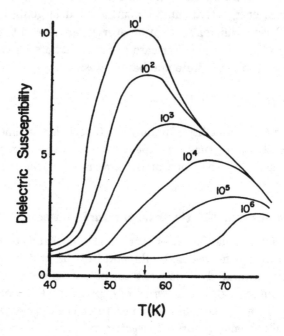

Fig. 6. The vector amplitude of the dielectric susceptibility of $K_{0.974}Li_{0.026}TaO_3$ as a function of temperature. The measuring frequency is shown above the relevant curve. This figure is taken from the work of Hochli.[73]

5.2. Distribution of Relaxation Times

The spin glasses show a broad spectrum of relaxation times below the glass transition, due to the hierarchical nature of the relaxational process as discussed above. Similar effects are seen in orientational dipolar glasses,[33,34] where we expect that a similar effect occurs due to the formation of local domains of polarized atoms. In Fig. 7 we show the complex dielectric response of $(KBr)_{0.5}(KCN)_{0.5}$ as a function of frequency and temperature.

This figure is rather messy, but if you trace the curves for various temperatures you will observe a peak in the loss tangent which seems to rapidly move to very low frequencies as one passes through a temperature, which we can loosely call a glass transition temperature, at about 20 K or so in the case of this material. Further, the width of the imaginary component ϵ_2 cannot be fit by a single Debye relaxation time. Thus, we see a divergence in the relaxation time and an anomalous width to the loss peak.

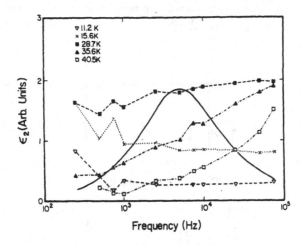

Fig. 7. The Fourier spectrum of the complex dielectric constant of $(KBr)_{0.5}(KCN)_{0.5}$ as a function of temperature. The solid curve is the prediction of the Debye relaxation model with a single relaxation time of $\tau = 3.18 \times 10^{-3}$ sec. This isn't the world's clearest figure, but you will notice that a really ragged looking distribution moves down from high frequencies (40.5 K) to low frequencies (11.2 K). Biophysicists do a much better job than this!

5.3. Elastic Moduli

In the elastic quadrupole systems the strain field is the predominant carrier of the interaction between the electric quadrupoles. Below the glass transition the dipoles are effectively frozen in to a given configuration. Since it is the strain field that carries the dipole interaction the effect is as if the strain field had become considerably larger below the glass transition temperature, or as if the elastic moduli had become very large. Another way to view this is to realize that the elastic moduli are the equivalent of the inverse of a mechanical susceptibility: a large elastic modulus corresponds to a small strain for a given stress, and hence a small susceptibility. Thus, in an elastic glass we would expect to see a *minimum* in the Young's or shear modulus at the glass transition, and rise in the modulus on either side of the transition. In fact, exactly this kind of behavior is observed in elastic glasses such as $(KBr)_{1-x}(KCN)_x$.[35] Figure 8 gives a plot of the shear modulus of an orientational glass as a function of temperature, and the loss decrement. Note the deep minimum in the shear modulus at the glass transition. We need to stress here that this deep minimum could have strong implications for a biological system, since at the elastic modulus minimum the system is most flexible. We should be ready for the proteins now!

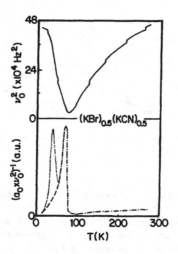

Fig. 8. The square of the resonance frequency of a torsional pendulum oscillator, which is proportional to the shear modulus, in the orientational glass $(KBr)_{0.5}(KCN)_{0.5}$. The top curve gives the square of the resonance frequency, and the bottom curve is the loss decrement. This figure is taken from the paper of Knorr *et al.*[35]

6. Spin Glasses and Proteins

Let us now try to draw the analogy between the spin glass described above and the globular protein, having educated ourselves in the above sections on "clean" spin-glass physics. As we stress repeatedly in this paper, one can only draw analogies to spin-glass. If you seek a rigorous connection with spin glasses, read no further and go back to your mathematics. Some of the buzz words used in spin glass physics have been applied to proteins. Let's see how well the analogy works, and if any insight into biological systems can be arrived at. We will cover the proteins in roughly the same order of subjects that the spin glass was developed.

6.1. Low Temperature Recombination Kinetics and Distributed Kinetics

Given the enormous range of physical aspects of proteins that can be studied, it is fair to ask what is an important problem, and what tools exist to solve the problem. For one of us (RHA), the motivation for examining spin-glass models for protein behavior stemmed from work at the University of Illinois, under the guidance of Hans Frauenfelder.[36–38] This was a simple experiment involving the observation of a chemical reaction in myoglobin at cryogenic temperatures, from 300 K to 4.2 K. We observed that below 200 K in a glycerol water glass the rate of the reaction, i.e., the recombination of carbon monoxide with iron, could not be characterized by a single rate constant but instead seemed to be due to a *spectrum* of rates.

That is, if a reactant has to surmount a *single, mono-energetic* barrier of height E_a then the rate k for the reaction should proceed as:

$$k = A \exp(-E_a/kT) \tag{18}$$

and the actual number of molecules $N(t)$ surviving after a time t is given by:

$$N(t) = N(o) \exp(-kt) \tag{19}$$

Note please (please!) that the deviations from an exponential are basically believed to be due to: (1) A *HETEROGENEOUS* distribution of occupied states and 2) The distribution in sites is *CONTINUOUS* and *NOT* discrete. This is a supposition at this point .... do the data stand up to these two critical requirements?

Given that a few of you out there believe that there really is a continuous distribution of activation energies the next question is: can we understand the functional form of the distribution? There have been three major attempts to understand the form of the distribution; one of them, by Stein,[39] directly used the spin-glass analogy. The other two come from rather different camps. The model of Bowne and Young[40] explicitly uses an adiabatic formulation to model the low temperature recombination process[41] and assumes anharmonic potential wells based upon the temperature-dependent

X-ray scattering data of Petsko and coworkers.[42] The third model, due to Agmon and Hopfield,[43] use a distribution arising from Franck-Condon factors and harmonic potential wells. In any event, all models do agree on one thing: that there is a distribution of activation energy barriers!

Figure 9 shows the data from one of the original papers plotted on a log $N(t)$ vs. log t plot. The data are obviously not single exponentials. Are the data multiple but discrete exponentials? Casual inspection of Fig. 9 indicates about 6 exponentials would be needed to fit the data, expanded data by Frauenfelder's group which covers 12 decades of time would need on the order of 12 exponentials, so as the number of exponentials goes to infinity we arrive at a continuous distribution of states. *We thus assume that the activation energies for recombination are given by a probability distribution g(E) for finding a molecule with activation energy E, or equivalently a distribution g(k) for finding a molecule with rebinding rate constant k.* The kinetics then become:

$$N(t) = N(o) \int g(k)e^{-kt}dk. \qquad (20)$$

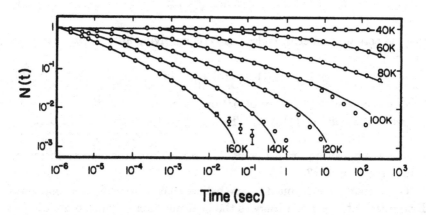

Fig. 9. A plot of the log of the number of unrecombined molecules $N(t)$ vs. log time in a myoglobin-CO photolysis experiment. The solid line is the best fit of the activation energy spectrum for myoglobin shown in Fig. 10.

Figure 10 shows the spectrum of activation energies $g(E)$ that we used to fit the data.

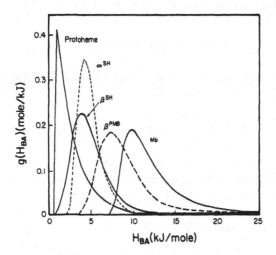

Fig. 10. A plot of the spectrum of activation energy $g(H_{ba})$ needed to fit low temperature recombination data for various heme proteins. The curve labeled Mb is the distribution needed for myoglobin.

Basically, this experiment revealed that at low temperatures the protein seemed to have a time-invariant and temperature invariant *spectrum* of activation energies. Other workers have proposed that the distribution of rates is due to internal dynamics within a single molecule,[44] or that a small number (4 or less) of exponentials can be used to fit the data,[45] but several experiments have ruled those possibilities out. If you bought the spin-glass analogy up to this point, but disagree with the continuous distribution explanation of our data, sorry to see you go. Acceptance of this point is rather critical for the following material.

Dan Stein was, we believe, the first to draw the analogy between the low temperature distribution of states and the predictions of the spin glass model.[39,46] In some sense what Stein did was to intuitively sense the physical similarities between the two systems and use an ansatz to map the distribution in energy states seen in one system over into another system. Thus, while what he did was not rigorous, it provided inspiration for several in the field. Stein simply noted the correspondence between spin glasses and orientational glasses, assumed a Gaussian distribution of activation energies arising from a Gaussian distribution of energies within most spin-glass models and used this reasonable approximation to fit the low temperature recombination data.

Stein used a Gaussian coupling constant between spins:

$$P(J_{ij}) = \frac{1}{\sqrt{2\pi J^2}} \exp(-(J_{ij}^2/2J^2)) \tag{21}$$

and then supposed that the result that the probability distribution of the metastable spin states of energy E is:

$$P(E) = \frac{1}{N\pi J^2} \exp(-(E - E_o)^2/NJ^2). \tag{22}$$

This Gaussian distribution in energy levels was then "frozen" in at some temperature T_f, presumably the glass transition temperature of the protein, to yield the final energy distribution:

$$D(E) = \exp(-(E - E_o)^2/NJ^2) \times \exp(-E/k_b T_f). \tag{23}$$

This distribution, "predicted" from spin-glass physics, was used to fit the recombination data of T-state Hb. Figure 10 is taken from Stein's paper and shows a comparison of a Gaussian distribution with the actual data. The fit is reasonably good, although the model of Bowne and Young[40] claimed to achieve substantially better χ^2. It isn't clear to us that the two models clash with one another; there probably are a variety of ways to express phenomena in complex systems, although a gaussian is quite an expected functional form for a random process. Indeed, the model of Agmon and Hopfield,[43] although it does not explicitly use any concepts from spin-glasses, also arrives at a distribution which is also Gaussian.

Actually, although Young and Bowne ruled that the simple Gaussian distribution of activation energy barriers did not fit the recombination data, recent results using the so-called "A" state infrared CO stretch bands, which are split in myoglobin, seem to reveal that the recombination kinetics on a *single* band are actually quite well fit by a Gaussian distribution![47] The story doesn't seem to be over yet. Probably the main point is simply that a straightforward application of the Edwards-Anderson spin-glass model to the barrier distribution in heme proteins gives a "reasonable" fit to the data.

6.2. Rare-Earth Luminescence and Conformational Disorder

As we stated earlier, two important ingredients of a spin glass are randomness and frustration. The question of frustration has not yet been clearly addressed in proteins: in principle the computer codes that calculate molecular dynamics could check for frustration in the protein structure since the potentials are believed to be accurate representations of the correct potentials.

One of us (RHA) carried out an experiment with a student to test for randomness in local protein structure,[48] with strong motivational pressure from Dan Stein. The idea was quite simple. Rare earth ions have long emission lifetimes which are sensitive in a known manner to their environment. Because of this sensitivity, it can be shown that the decay rate Γ_{if} of an excited rare earth ion from an excited state i to a final state f in the presence of a monopole charge at a distance R is:

$$\Gamma_{if}(R) = \frac{A}{R^6} + \frac{B}{R^{12}} + \frac{C}{R^{16}} \tag{24}$$

where A, B and C are factors which contain factors arising from the detailed atomic parameters.

The distribution of the monopole position R as a function of different protein conformers is assumed as usual to be Gaussian:

$$P(R) = \frac{\exp[-(R - R_o)^2/2\sigma^2]}{(2\pi\sigma^2)^{1/2}} . \tag{25}$$

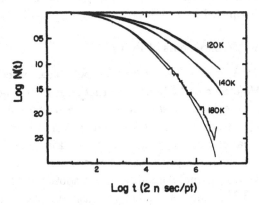

Fig. 11. The terbium luminescence vs. time for the rare earth Tb^{3+} bound to the protein calmodulin. The solid line is the best fit of the equation in the text to the data.

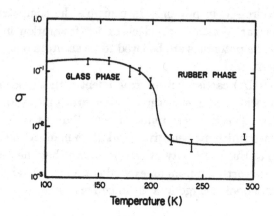

Fig. 12. The apparent width of the Gaussian distribution of distances taken from Eq. (24) and the fit to the data in Fig. 11 as a function of temperature.

Figure 11 shows how this Gaussian distribution of distances can fit the data for terbium luminescence in the protein calmodulin at several cryogenic temperatures, and Fig. 12 gives a semi-log plot of the width of the distribution as a function of temperature in a glycerol/water solvent, which becomes glassy at approximately 200 K. Note that at temperatures above the glass temperature of the solvent (and, apparently, the protein) the protein structure rapidly averages to give, effectively, a very narrow "width". Together, these experiments seem to argue for both the presence of a glass transition in a protein and the presence of a distribution of conformational states in the glassy protein.

6.3. Specific Heat Measurements of Proteins

One of us (RHA) was a co-editor of a conference proceedings: *Protein Structure: Molecular and Electronic Reactivity*. In this book there are a number of experimental articles showing how a protein shows glass-like properties at temperatures below the glass transition. The most direct evidence is from the paper of Gol'danskii et al.[49] and Finegold,[50] who studied the specific heat of hydrated proteins at low temperatures. The data, as Gol'danskii points out, is unfortunately rather spotty in the temperature range of interest.[50-52] Although the early measurements by Finegold and colleagues attempted to fit the low temperature specific heats to either a varying-dimensionality model or to computer simulations, the "modern" view of this data is to fit C_p to a semi-empirical formula:

$$C_p(T) = C_1 T + C_2 T^3 + C_E(\Theta_E/T) \tag{26}$$

where C_1 is the specific contribution due to amorphous states as we discussed above in the spin glass section, C_2 is a T^3 contribution empirically tied to a Debye relaxation process in 3 dimensions, and C_E is the Einstein coefficient contribution. With so many variables, it is not too surprising that the above equation fits the rather limited data rather well, as we show in Fig. 13. The complexity of the equation should not detract from the fact that the interesting term C_1 is dominant at low temperatures.

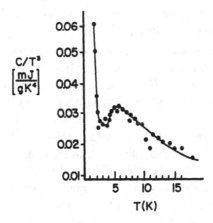

Fig. 13. The specific heat C/T^3 vs. temperature for bovine serum albumin in a desiccated crystalline state. From the work of Goldanskii *et al.*[49]

The conclusion that one can draw from this is that at low temperatures proteins do indeed seem to resemble disordered glasses in analogy to the spin glass or the orientational glasses in at least one respect: the dominant linear specific heat dependence. Anderson *et al.*[23] and Phillips[24] would say that the structure is characterized by a large number of 2-level tunneling states, which is closely related to the conformational states we have been talking about. Probably more detailed work should be done in this area. The complexity of the equation used to fit the data causes some discomfort, and we wonder if some of the more recent ideas concerning phonons on disordered lattices[53] might be applicable here. More experimental and theoretical work to be done here, but we have to confess it is not exactly the kind of thing to make the blood hot. Best left to graduate students who like to keep tidy and neat desks. The low temperature specific heat of proteins doesn't give too much information about what happens up at higher temperatures!

6.4. Dielectric Susceptibility

Susceptibility measurements give information about the conformational flexibility of the protein. There has been very little systematic work on the dielectric susceptibility of proteins, in glaring contrast to the lavish effort expended in spin glasses on magnetic susceptibility. This is quite unfortunate, since the characteristic cusp in the susceptibility and the frequency dependence of the cusp have been instrumental in identifying several of the properties of spin-glass transition.

Parak and his coworkers have done a series of experiments of dielectric relaxation rates in metmyoglobin crystals,[54] but unfortunately the high frequency they chose to use (10 GHz) meant that they predominantly studied the relaxation rate of *water* in the crystal rather than the protein structure. This experiment did show that the so-called bound water on the surface of a globular protein does not abruptly freeze at the bulk freezing point of the solvent water but instead gradually slows in its relaxation rate. However, this experiment, while important in understanding the relationship between the Mossbauer relaxation rate and the bound water relaxation rate, does not directly address the issues raised in this review.

There have been lower frequency dielectric relaxation work done by the Dielectrics Group at the University of London[55] which is related to some of the issues we want to address here, but they have not done a systematic study of the relaxation rate as a function of temperature nor is their model couched in spin-glass language (this is not of course a strike against them!). However, in the paper by Dissado,[55] there is a clear reference to the existence of distributions in relaxation times and conformational heterogeneity.

Over the years, one of us (RHA) has worried about this lack of a clear experimental attack on the temperature dependence of the dielectric response of proteins. I have tried to induce undergraduates and graduate students to mount a vigorous attack, but I have failed miserably and the efforts have been at best half-hearted. I would like to blame the students for their lack of imagination — and do to some extent — but the main fault lies with the adviser's failure to communicate to the students the importance of the measurement. I was going to plot the results of some experiments that we did on the dielectric susceptibility of a concentrated hemoglobin solution as a function of temperature. Even at the highest concentrations of proteins *in solution* that you can easily obtain (on the order of 20 milligrams/ml) the protein is still only one part in fifty of the total mass. Thus, any dielectric experiment involving proteins in solution must somehow remove the dielectric constant of the solvent. For what it is worth, the data that we have do show the presence of a *cusp* in the dielectric constant of the protein followed by a decrease of the dielectric constant as the temperature is lowered past the glass transition point of the solvent. The analogy to the susceptibility of the spin glass is clear, but for lack of a systematic attack on the problem the data can only remain suggestive at present. It is clear what has to be done: the dielectric response of the protein has to measured as a function of frequency and temperature, and the peak position of the cusp extrapolated to zero frequency to establish the glass freezing temperature. So simple, maybe somebody will do the damn experiment before we do.

6.5. Elastic Moduli

If the protein is truly a form of a orientational glass where the coupling between electric multipoles is via the strain field then it should show the same variation of elastic moduli with temperature as the simpler elastic orientational glasses.

The cleanest experimental verification is probably the beautiful paper by Morozov and Gevorkian.[56] In this paper the authors made solid samples of a variety of proteins, both crystalline and amorphous, and studied the viscoelastic properties of the samples as a function of temperature from 100–300 K. We don't have space in this review to discuss the effects of differing hydration level on the elastic properties of the proteins, which is truly interesting and important. Rather, let us consider the real component of the Young's modulus E as a function of temperature for a protein crystal saturated with water, as seen in Fig. 14. Note that as predicted the Young's modulus *increases* with decreasing temperature as does an elastic quadrupole glass, and that the increase is quite significant at low temperatures.

Unfortunately, the expected rise in the Young's modulus at higher temperatures is not seen here, and indeed it is problematical if significant

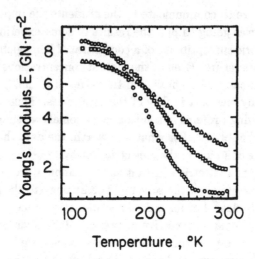

Fig. 14. The Young's modulus E of a tetragonal lysozyme crystal along the [001] direction as a function of temperature. The relative humidity of the sample was 94%. This figure is taken from Ref. 56.

increases in the temperature could be achieved without denaturing the protein. Further, because of the broadness of the transition it is difficult to ascribe a phase transition temperature to the samples. The lesson here, as far as it can be ascertained, is that the protein does indeed seem to show the elastic moduli variation expected of an elastic dipole glass, and that the variations in the Young's modulus are quite significant, indicating that some biological significance may also be ascribed to the glass transition. We like the last sentence of the Morozov and and Gerorkian paper, taken somewhat out of context: "This is an example of physics which biology uses to make proteins function".

6.6. Passage Through the Glass Transition: The Kinetic Signatures

A dynamic picture of the glassy state of a protein at low temperatures can be found by using the technique of flash photolysis, as we discussed in the earlier part of this section. Glass transitions can be identified by rather dramatic changes in both dynamic and thermodynamic quantities.

The time course of rebinding gives important information about the dynamics of the protein myoglobin. In particular, even if the spin-glass analogy allowed us to confidently predict that the protein had many conformational states, if the myoglobin freely sampled all of its states on a time scale much faster than the mean recombination time then the recombination will be a simple exponential, as is indeed observed at room temperature. Now, as the protein is cooled we might hope that a glass transition will occur, as happens in spin glasses and in orientational glasses. Below the glass transition temperature the system will no longer be ergodic: the dynamical divergence of the relaxation times will keep various protein molecules in various states for effectively infinitely long times. We should then see deviations from simple exponential recombination. Depending on the solvent, the kinetics change from the temperature invariant distribution of rates at low temperatures to a quasi-narrow single rate at high temperatures. This change occurs over a narrow temperature range that seems to be linked to the glass transition of the solvent,[57] a fact that has given rise to the concept of a "slaved glass" transition in the protein.

Recently, some exciting work by Hans Frauenfelder and his colleagues has resulted in direct measurements of a nuclear coordinate in the vicinity of the "slaved" glass transition.[57] Frauenfelder has exploited the fact that the CO stretch band of iron-ligated CO in most heme proteins is split into a number of sub-bands,[61] called "A" states by Frauenfelder due to his historical identification of the bound CO with the label "A". These bands are sensitive to many external parameters, including temperature and pressure.

Observation of the ratio of the A_o to A_1 as a function of temperature at a static atmospheric pressure revealed that the temperature dependence of the ratio of the states stopped at the glass transition temperature of the solvent. Thus, the ability of the conformational distribution to adjust to temperature seems to halt at the external glass transition temperature, as we would expect from a number of the experiments discussed above. Of greater interest is the question: what is the rate at which the protein is able to approach equilibrium as the glass transition is approached from above?

Actually, this is a rather deep question, especially when recast into some of the language we used to discuss the transition in a spin glass. That is, is there a true thermodynamic phase transition underlying the kinetic slow down of the glass transition, or do we merely observe a thermally driven falling out of equilibrium? Frauenfelder was able to look at conformational relaxation by cooling the protein under hydrostatic pressure to a given temperature T and then suddenly releasing the pressure. The ratio of the A states relaxes to a new value appropriate to the equilibrium value at the lower pressure. It was observed that as the glass transition was approached, the relaxation kinetics were non-exponential and highly temperature dependent. A loose analogy to this experiment in spin-glass lore would be to suddenly decrease the magnetic field on the spin-glass sample at some temperature T and observe the relaxation of the magnetization, although the closer analogy of actually changing the hydrostatic pressure on the sample in a fixed magnetic field has, to my knowledge, not been done.

The interesting parameter that Frauenfelder *et al.* measure here is the temperature dependence of the relaxation process near the glass transition. Frauenfelder *et al.* choose to fit the relaxation to a Bassler-Zwanzig function:

$$k(T) = k_o \exp[-(T_o/T)]^2 \tag{27}$$

rather than the Vogel-Tamman-Fulcher relation:

$$k(T) = k_o \exp[-E/(k_b(T - T_o))] \tag{28}$$

to fit the data. The question of which function to use maps back to the question of whether the glass transition is some sort of hydrodynamic arrest or has underneath it some sort of a phase transition. A clear summary of the differences between these two pictures can be found in the paper of Bassler.[59] The argument we have is that the VTF relation has a singularity at the critical temperature T_o rather then the continuous relationship of the stretched exponential. Both curves give adequate fits to the relaxation data, but the extrapolation into lower temperatures is of course completely different. Unfortunately, *both* equations give impressively good fits to data over 9 orders of magnitude![58]

In the case of the glass relaxation work of Frauenfelder, the relaxation of the ratio of the peaks was given by a power law:

$$\Phi_r(t) = [1 + k_r(T)t]^{-n} \tag{29}$$

and fits were done with n as a variable and k_r was fit to the hydrodynamic law.[60] Figure 15 shows the fits that were obtained using this model. It is clear that proteins at present offer no clear test of the troubling question concerning the possibility that phase transitions do or do not exist in glasses (in general) or proteins (in particular). These are excellent experiments, and intriguing results, but the fundamental questions are still not tested.

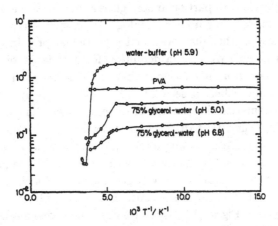

Fig. 15. A plot of the ratio of the integrated area of the infrared CO stretch bands A_0/A_1 as a function of time. For details consult Ref. 60.

It does seem clear, however, that the protein shows kinetic aspects that spin glasses and glasses show: significant slowing down near some sort of a fixed temperature.

6.7. Hierarchical States

Unfortunately, the discussion in this section will be even more murky than the one in the spin glass section, if that is possible. The one paper that made a real stab at using these terms to explain some experimental aspects of protein behavior was written by Frauenfelder's group at Illinois.[61]

In this paper the authors attempted a synthesis of experiments from a large range of physical techniques in order to show that the structural relaxation of the protein myoglobin could be characterized as a hierarchal diffusion through connected states. The main thrust of the paper was to show that the sets of states through which the protein relaxed after photolysis could be characterized within a hierarchical scheme of progressively large motions.

Frauenfelder coined the phrase Conformational Substates, abbreviated as CS, to describe these hierarchically connected substates. Thus, CS0 would be the set of substates closest to the iron atom, while CS4 would be the substates associated with conformationally distinct protein substates that might have different surface configurations. It is a little strained to connect these CS levels with particular spin glass states since the 3-dimensional topology of a protein has no easy analog with a spin glass. The analogy should be instead to the "distance" that a particular protein conformation is from another one. We used a definition for a distance of one particular spin glass state from another one that made sense, yet no one (except maybe Karplus and Elber, as seen later) has really come up with some sort of a way to systematize the concept of distance clearly in proteins — and we must keep in mind that distance may not have the intuitive meaning that we are used to. That is, two protein configurations may be rather close to one another as viewed by X-ray diffraction yet the folding path between the two conformations could be very large. This lack of a clear definition of distance in the protein systems will cause us much grief later in the dreaded ultrametric section. Figure 16, taken from Ref. 62 gives in schematic form the essence of Frauenfelder's ideas.

The next step removes us far from any idea in spin glasses and separates the physicists from the biophysicists. No one has ever spoken about a functionally important motion in a spin glass, and probably would be driven from the high holy temple of condensed matter physics if they did. However, in a protein like myoglobin there really exists two different sets of conformational substates: those associated with no bound ligand (for example) and those that are associated with the bound ligand. The connection between these two sets is via what Frauenfelder called a *functionally important motion*, or a FIM. We can't think of an analogy in the spin glass system that would be physically realizable. Frauenfelder viewed the recombination process as consisting of relaxation between the CS's and lateral movements over via the FIM's.

The paper was possibly flawed by interpreting the shift in a near-infrared charge transfer band at 760 K. The evolution of the maximum of this band vs. recombination at low temperatures (less than 180 K) was interpreted as evidence for conformational flow of the structure within the CS_2. Several workers[63,64] have subsequently pointed out that in fact what was occurring was a form of reactive hole burning. That is, each different conformational state of the protein has a particular band near 760 nm, and as recombination proceeds the band appears to shift as the long-wavelength sub-bands combine. The effect of this hole burning is to make the maximum of the band appear to "move", although no conformational relaxation is occurring. Thus, rather than relaxing the substates in CS_2 are actually temperature invariant, along the lines of a glass transition.

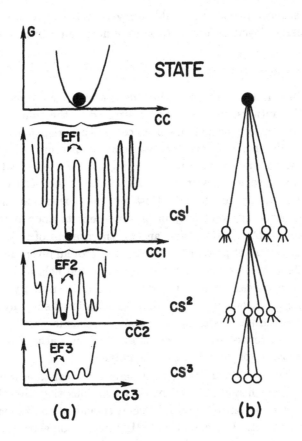

Fig. 16. A conceptual sketch of the hierarchical nesting of conformational states in a protein. The conformational levels are denoted by the subscript, with increasing number indicating ever smaller displacements from the proposed global energy minimum. From the work of Frauenfelder *et al.*[62]

In our view this flaw in no way invalidates the basic ideas of Frauenfelder, in fact, the existence of hole-burning would seem to give strong support to the idea of a distribution of conformationally distinct substates! It would be interesting in our opinion to continue pushing on the hierarchical concept especially in the light of kinetics theories to determine the actual rate of flow of the substates in the protein and verify that hierarchical complexity serves as a kinetic bottleneck in the process.

In general, to us the analogy of hierarchal structures in the protein seems apt, since the compact folding of the protein in what seems to be a directed sequential manner would imply that the structure must flow through different layers of organization to undergo an arbitrary relaxation.

6.8. Ultrametricity

This is the most dangerous section (dangerous to the professional reputation of the authors, that is). As we stressed in the spin-glass section, ultrametricity is a very mathematical concept with a crisp definition. Questions arise as to whether a real 3-D spin glass is ultrametric, so things in the protein arena will be much worse. The ghosts of theoretical physicists crowd into the office and glower over our shoulder at the LCD screen, muttering *bullshit!* under their breath. However, we believe that the question is important, since ultrametricity in proteins reflects upon both how the protein is folded into its structure, and on the overlap of adjacent structures. We cannot as yet predict the folding of a protein given the sequence, so even as abstruse a concept as ultrametricity could help us to codify the problem. Perhaps a clean way to put it is this: if ultrametricity is operative in protein conformation space, then the conformational substates of a protein do not arise from a "kicking, screaming stochastic walk" as Gregorio Weber has characterized it, but instead evolve from paths determined by the previous history of the folding of the polymer.

We'll put our cards on the table right here: it is clear to us from the fact that proteins form history-dependent conformations that the ultrametric idea is of great importance[65] and full credit should go to Frauenfelder for pushing this concept. But, one of us (RHA) has to somehow satisfy those lemma-oriented ghosts howling in my office: what to do?

Frauenfelder and his coworkers in the paper cited above for hierarchies also were bold enough to claim that the protein space could also be ultrametric,[61] and have also discussed the concept in several review papers.[66,67] The concept of ultrametricity is of very little use unless there exists a very crisp definition of distance by which to address the ultrametricity question. An attempt to come up with a workable definition of distance between protein conformations was made in a paper by by Karplus and Elber,[68] on the basis of computer simulations. They decided that ultrametricity was of little use, but there was a flaw in the paper as written which makes the issue not so clearly dead.

In their paper, a 300 picosecond simulation of myoglobin structural relaxation was done. A set of randomly chosen structures consistent with

X-ray crystallography were allowed to relax over this time range, and the root-mean-square differences between the structures was compared before and after relaxation to determine if two nearby initial structures converged to a common structure or diverged to separate structures that are separated by an energy barrier, as in Fig. 17. In essence, this is equivalent to the test within spin-glass physics as to the presence of nearly iso-energetic ground substates separated by energy barriers.

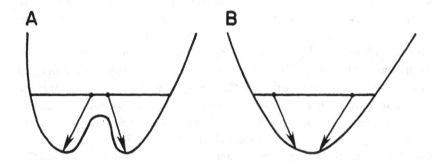

Fig. 17. Two possibilities for structural relaxation in a computer experiment. If two initial conformations represent two separate minima, then an energy barrier will separate them, as in figure A. If they belong to the same minimum energy, then they will converge as in figure B. Taken form the work of Elber and Karplus.[68]

The basic concept of a glass-like structural space was verified by the observation that the configuration space seemed to be made up of many minima with small energy differences. A test was then made to see if the space was ultrametric. The "distance" between different stable configurations K and K' was defined by:

$$D_{K,K'} = \sum_{i,j} \Delta_{ij}(K, K') = \sum_{i,j} R_{ij}(K) - R_{ij}(K') \qquad (30)$$

where $R_{ij}(K)$ is the distance between the amino acid units i and j in configuration K. Similar structures have small $D_{K,K'}$, while dissimilar structures have large $D_{K,K'}$.

Let us now ask how we can use this matrix to test for ultrametricity. Suppose we look at, say, N=50 different stable conformations (denoted by K running from 1 to 50). If two structures are closer than some given distance apart then we can define them to be in the same cluster. Ultrametricity occurs if the grouping of similar structures via the distance matrix results in *disjoint* clusters. Of course, the size of the clusters is dependent on the amount of overlap defined for similarity. What does ultrametricity mean physically for the protein substates? It means that protein structures *within a cluster* evolve like a species, retaining their identity and not blurring into a structure which could arose from another disjoint cluster.

Now, in the paper by Elber and Karplus one of us (CC) noticed a mistake in the logic. As the paper was written, the authors confused "distance" with "overlap". Thus, their statement that "There is a rather sharp transition between the range $(D_{K,K'} \geq 1.5 \text{ Å})$ when all structures are disjoint, and the range $(0 \leq D_{K,K'} \leq 1 \text{ Å})$ when all the structures belong to the same cluster."[68] This statement makes no sense as we hope the above discussion made clear: if all the structures have less than 1 Å difference, there surely will be none with a difference greater than 1 Å! In a personal communication with Dr. Elber we received the clarification that the offending sentence should have read: "All the clusters form disjoint clusters for $(0 \leq D_{K,K'} \leq 1 \text{ Å})$ and a single cluster at $0 \leq D_{K,K'} \leq 1.5 \text{ Å}$", that is, there are no structures greater than 1.5 Å apart from each other. The corrected statement is the *logical inverse* of the original statement. This clarification makes the ultrametric nature of the conformational substates not so useless a concept as it appeared!

To see why that is true, let's refer to some data that Dr. Elber sent us of the actual distance matrix. Elber and Karplus picked out, presumably at random, 28 converged conformational substates after 300 ps of relaxation. The matrix elements $D_{K,K'}$ were evaluated for all the possible combinations and scored a 1 if $D_{K,K'}$ was LESS than some value, and a 0 if it were GREATER than some value. For a value less than some small number, such as 0.5 Å, we expect only the unit matrix and this is indeed seen in Fig. 18. As we let the distance cut-off increase, we do begin to obtain disjoint groups, as for example is seen in at $D_{K,K'} \leq 1.5 \text{ Å}$. There is a problem in the lower right hand corner: note that groups 25 and 26 are less than 1.5 Å apart, as are 26 and 28. However, 26 and 28 do not belong to the same cluster. Hence the system is not rigorously ultrametric. For distances greater than 1.5 Å we get one large group ball. Should Elber and Karplus have been in such a rush to throw out ultrametricity? We feel not. The fundamental ultrametric nature of the grouping is actually tous quite impressive, minus a few problems.

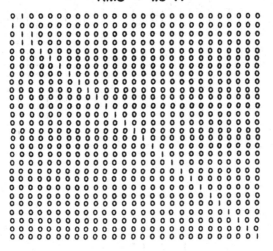

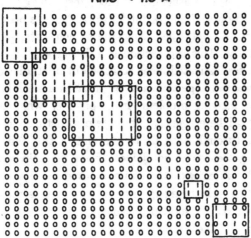

Fig. 18. The distance matrix $D_{K,K'}$ as a function of cutoff distance, for three different cutoff distances. "RMS" refers to the distance between different conformations of the protein. The conformations, 20 in all, are listed in order of increasing energy. See the text for details. This data was kindly provided by Dr. Elber.

RMS < 2.0 Å

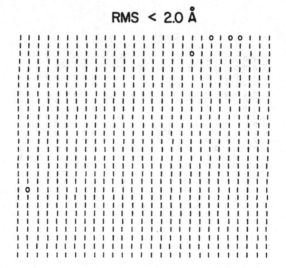

Fig. 18. (*Continued*)

Look folks, as we stated in the introduction our goal in biophysics is not to prove rigorously mathematical theorems! Our goal is to take ideas from some of the powerful and beautiful work that has been done in "clean systems" and try to apply these concepts to help categorize the complex molecules we study. This last section is a fitting way to close this paper: one of the most sophisticated ideas to arise out of studies on spin glasses, ultrametricity, was we hope shown to be *possibly* of some help in attacking one of the most nagging problems in biophysics: what is the nature of the conformational heterogeneity of protein structure. For those of you who have stuck through with us to the end of this tangled review, we hope you have gotten some good ideas from the spin glass analogy for EXPERIMENTS which will help us understand the dynamics of complex biopolymers. With luck these experiments will reveal some of the fantastic physics that must still be hidden in biology, and will give insight into the biological function of these molecules.

References

[1] R. Bone, J. L. Silen and D. A. Agard, *Nature* **339**, 191 (1989).
[2] J. Wong and C. A. Angell, *Glass: Structure by Spectroscopy* (Marcel-Dekker, NY, 1976).

[3] *Dynamic Aspects of Structural Change in Liquids and Glasses*, ed. C. A. Angell and M. Goldstein, *Ann. N.Y. Acad. Science* **484** (1986).

[4] M. A. Ruderman and C. Kittel, *Phys. Rev.* **96**, 99 (1954).

[5] T. Kasuya, *Prog. Theor. Phys.* **16**, 45 (1956).

[6] K. Yosida, *Phys. Rev.* **106**, 893 (1957).

[7] S. Miyashita, in *Topological Disorder in Condensed Matter*, ed. F. Yonezawa and T. Ninomiya (Springer-Verlag, New York, 1983), p. 191.

[8] K. Binder and A. Young, *Rev. Mod. Phys.* **58**, 846 (1986).

[9] D. S. Fisher, G. M. Grinstein and A. Khurana, *Physics Today* **14**, 56 (1988).

[10] G. Toulouse, *Commun. Phys.* **2**, 115 (1977).

[11] D. Sherrington and S. Kirkpatrick, *Phys. Rev. Lett.* **35**, 1972 (1975).

[12] S. F. Edwards and P. W. Anderson, *J. Phys.* **F5**, 965 (1975).

[13] F. Luty and Ortiz-lopez, *Phys. Rev. Lett.* **50**, 1289 (1983).

[14] E. R. Grannan, M. Randeria and J. P. Sethna, *Phys. Rev. Lett.* **60**, 1402 (1988).

[15] R. Pirc, B. Tadic and R. Bline, *Z. Phys.* **B61**, 69 (1985).

[16] C. Cantor and P. Schimmel, *Biophysical Chemistry, Vol. III* (W. H. Freeman, New York, 1980), p. 1041.

[17] C. Kittel, *Introduction to Solid State Physics*, (John Wiley, New York).

[18] C. A. M. Mulder, A. M. van Duyneveldt and J. A. Mydosh, *Phys. Rev.* **B23**, 1384 (1981).

[19] J. Ferré, M. Ayadi, R. V. Chamberlin, R. Orbach, and N. Bontemps, *J. Magn. Mater.* **54–57**, 211 (1986).

[20] W. Kauzmann, *Chem. Rev.* **43**, 219 (1948).

[21] D. L. Stein, *The Sciences, Sept.-Oct.*, **22** (1988).

[22] J. Jackle, *Rep. Prog. Phys.* **49**, 171 (1986).

[23] P. W. Anderson, B. I. Halperin and C. M. Varma, *Phil. Mag.* **25**, 1 (1971).

[24] W. A. Phillips, *J. Low Temp. Phys.* **7**, 351 (1991).

[25] H. G. B. Casimir and F. K. Du Pre, *Physica* **5**, 507 (1938).

[26] G. Parisi, *Phys. Rev. Lett.* **53**, 1754 (1979).

[27] M. V. Feigelman and L. B. Ioffe, *J. de Physique Lettres* **45**, 475 (1984).

[28] D. A. Huse and D. S. Fisher, *J. Phys. A: Math. Gen.* **20**, L997 (1987).

[29] R. G. Palmer, D. L. Stein, E. Abrahams and P. W. Anderson, *Phys. Rev. Lett.* **53**, 958 (1984).

[30] R. Rammal, G. Toulouse and M. A. Virasoro, *Rev. Mod. Phys.* **58**, 765 (1986)

[31] A. Ogielski and D. L. Stein, *Phys. Rev. Lett.* **55**, 1634 (1985).

[32] U. T. Hochli, *Phys. Rev. Lett.* **48**, 1494 (1982).

[33] S. Bhattacharya, S. R. Nagel, L. Fleishman and S. Susman, *Phys. Rev. Lett.* **48**, 1267 (1982).

[34] J. J. De Yoreo, M. Meissner, R. O. Pohl, J. M. Rowe, J. J. Rush, and S. Susman, *Phys. Rev. Lett.* **51**, 1050 (1983).

[35] K. Knorr, U. G. Volkmann, and A. Loidl, *Phys. Rev. Lett.* **57**, 2544 (1986).

[36] R. Austin, K. Beeson, L. Eisenstein, H. Frauenfelder, I. C. Gunsalus, and V. Marshall, *Science* **181**, 541 (1973).

[37] R. Austin, K. Beeson, L. Eisenstein, H. Frauenfelder, I. C. Gunsalus, and V. Marshall, *Phys. Rev. Lett.* **32**, 403 (1974).

[38] R. Austin, K. Beeson, L. Eisenstein, H. Frauenfelder, I. C. Gunsalus, and V. Marshall, *Biochemistry* **14**, 5255 (1975).

[39] D. Stein, *Proc. Natl. Acad. Sci. USA* **82**, 3670 (1985).

[40] R. D. Young and S. F. Bowne, *J. Chem. Phys.* **81**, 3730 (1984).

[41] H. Frauenfelder and P. Wolynes, *Science* **229**, 337 (1985).

[42] H. Frauenfelder, G. A. Petsko and D. Tsernoglou, *Nature* **280**, 558 (1979).

[43] N. Agmon and J. J. Hopfield, *J. Chem. Phys.* **78**, 6947 (1983).

[44] W. Bialek and J. N. Onuchic, *Proc. Natl. Acad. Sci. USA* **85**, 5908 (1988).

[45] L. Powers and W. Blumberg, *Biophysical J.* **54**, 181 (1988).

[46] D. L. Stein, in *Protein Structure: Molecular and Electronic Reactivity* (Springer-Verlag, New York, 1987), p. 85.

[47] Joel Berendzen, personal communication.

[48] R. H. Austin, D. L. Stein and J. Wang, *Proc. Natl. Acad. Sci. USA* **84**, 1561 (1987).

[49] V. I. Gol'danskii, Y. F. Krupyanskii and V. N. Fleurov, in *Protein Structure: Molecular and Electronic Reactivity*, ed. R. Austin *et al.* (Springer-Verlag, New York, 1987), p. 95.

[50] B. Fanconi and L. Finegold, *Science* **190**, 458 (1975).

[51] L. Finegold and J. L. Cude, *Nature* **238**, 38 (1972).

[52] B. I. Verkin, V. Suharevskii, Y. Telezhenko, A. V. Alapina, and N. Y. Vorob'eva, *Fizika Nizk. Temp* **3**, 252 (1977).

[53] I. Webman and G. S. Grest, *Phys. Rev.* **B31**, 1689 (1985).

[54] G. P. Singh, F. Parak, S. Hunklinger and K. Dransfeld, *Phys. Rev. Lett.* **47**, 685 (1981).

[55] L. Dissado, *Protein Structure: Molecular and Electronic Reactivity* (Springer-Verlag, New York, 1987), p. 47.

[56] V. N. Morozov and S. G. Gevorkian, *Biopolymers* **24**, 1785 (1985).

[57] I. E. T. Iben, D. Braunstein, W. Doster, H. Frauenfelder, M. K. Hong, J. B. Johnson, S. Luck, P. Ormos, A. Shulte, P. J. Steinbach, A. H. Xie, and R. D. Young, *Phys. Rev. Lett.* **62**, 1916 (1989).

[58] Y. Jeong, S. R. Nagel and S. Bhattacharya, *Phys. Rev.* **A34**, 602 (1986).

[59] H. Bassler, *Phys. Rev. Lett.* **58**, 766 (1987).

[60] A. Ansari, J. Berendzen, D. Braunstein, B. Cowen, H. Frauenfelder, M. K. Hong, I. E. T. Iben, J. B. Johnson, P. Ormos, T. B. Sauke, R. Scholl, A. Schulte, P. J. Steinbach, J. Vittitow, and R. D. Young, *Biophys. Chem.* **26**, 337 (1987).

[61] A. Ansari, J. Berendzen, S. F. Bowen, H. Frauenfelder, I. E. T. Iben, T. B. Sauke, E. Shyamsunder, and R. D. Young, *Proc. Natl. Acad. Sci. USA* **82**, 5000 (1985).

[62] H. Frauenfelder, in *Protein Structure: Molecular and Electronic Reactivity* (Springer-Verlag, New York, 1987), p. 245.

[63] N. Agmon, *Biochemistry* **27**, 3507 (1988).

[64] B. C. Campbell, M. R. Chance and J. M. Friedman, *Science* **38**, 373 (1987).

[65] J. Maddox, *Nature* **324**, 205 (1986).

[66] H. Frauenfelder, in *Amorphous and Liquid Materials*, Proceedings NATO Summer School, ed. E. Luscher, G. Fritsch, and G. Jacucci (Martinus Nijhoff, 1987), pp. 3–18.

[67] H. Frauenfelder, in *Physics in Living Matter*, Lecture Notes in Physics 284 (Springer-Verlag, 1987), pp. 1–14.

[68] R. Elber and M. Karplus, *Science* **235**, 318 (1987).

[69] H. Maletta and W. Felsch, *Phys. Rev.* **B20**, 1245 (1979).

[70] G. E. Brodale, R. A. Fisher, W. E. Fogle, N. E. Philiips, and J. van Curen, *J. Magn. Matter* **31–34**, 1331 (1983).

[71] L. E. Wenger, in *Proceedings of the Heidelberg Colloquium on Spin Glasses*, ed. J. L. van Hemmen and I. Morgenstern, Lecture Notes in Physics Vol. 192 (Springer, Berlin), p. 60.

[72] J. J. Ferre, J. Rajchenbach and H. Maletta, *J. Appl. Phys.* **52**, 1697 (1981).

[73] U. T. Hochli, *Phys. Rev. Lett.* **48**, 1494 (1982).

Spin Glass Ideas and the Protein Folding Problems

Peter G. Wolynes

Chemistry Department and Beckman Institute,
University of Illinois, Urbana, IL 61801, USA

We do not yet understand how protein sequence is translated into three-dimensional protein structure. This article explains how spin glass ideas may be of relevance in solving this "protein folding problem". Spin glass phase transitions may influence the rate of folding. We discuss this using a statistical protein model. Sophisticated spin glass theories of heteropolymer collapse are also discussed. Connections with Hopfield associative memory models may give predictive schemes for solving the problem.

1. Introduction

The key to our current understanding of molecular biology has been the recognition of the importance of information transfer in living things. A good deal of this information transfer is easy to understand as simple recording of the same information. The one-dimensional sequence of a DNA molecule determines the sequence of an RNA molecule which further determines the sequence of a protein molecule, word for word. In addition, we understand how one part of a gene may turn on or turn off another part of a gene, much in the manner of instructions of a Turing machine tape. A real gap in our understanding of information transfer, however, is apparent when we try to explain how the sequence of a protein determines its three-dimensional structure. It is the three-dimensional structure of a protein molecule that determines its adequacy for function. Only a correctly three-dimensionally structured protein molecule can act as an enzyme, or as an information carrying signal to other molecules. This gap must, therefore, be filled if we are truly to say that we understand biological function and its connection to genetics.

This "protein folding problem", i.e., how the one-dimensional sequence information is translated into a three-dimensional structure, has been singularly proof to the traditional methods of molecular biology and theoretical chemistry and physics. Perhaps this is because of the unfamiliar nature of one-dimensional to three-dimensional information transcription; the information transfer is more like a parallel computation than a trivial translation of a message. The solution of this problem would be of great practical importance. The technology for obtaining one-dimensional sequence information is quite advanced and is continuing to advance rapidly. If the project to sequence the human genome is carried out, the amount of one-dimensional sequence information available to scientists will be exponentially larger than we now have. On the other hand, the techniques for experimentally determining three-dimensional structure directly are advancing, but not nearly so rapidly. Thus it seems, as a practical matter, that we will have a tremendous backlog in the pipeline from the one-dimensional sequence information to three-dimensional structural information that we need in order to understand function.

Unlike many problems in molecular biology, this information transfer problem involved in protein folding may well be solvable by the traditional methods and techniques of physics and chemistry. For many proteins it is clear that there is no need for special biological machinery to accomplish folding.[1] (This may not be entirely universal, however. Recently evidence for folding catalysis has been uncovered in special cases.[2]) Proteins can be unfolded in the test tube and refolded simply by changing the physical and chemical environment. Thus under some circumstances, the correctly folded protein is a stable equilibrium structure despite the complexity of the sequence information and the seeming complexity of the resultant structures.

In the last decades, as illustrated in this book, a variety of complex structures and their stability have been investigated from the viewpoint of the theory of spin glasses. Spin glasses have been the paradigm for reconciling diversity and stability. Ideas from the theory of spin glasses have been used in understanding the complexity of neural information processing,[3] the immune response[4] and early evolution.[5] It is perhaps then inevitable that the ideas from the theory of spin glasses would be brought to bear on this problem of protein folding. In this review I will discuss attempts to use spin glass theory to understand the thermodynamics and dynamics of protein folding in a qualitative way and some recent attempts at using spin glass ideas to look at the practical problems of inferring three-dimensional structure from one-dimensional sequence. To understand the issues in this area, it is important for the reader to know some experimental facts about the structure of proteins and about experiments carried out to study the folding process.

In the next section I will give a very brief review of these areas for non-specialists. Following this, I will explain various approaches to the thermodynamics of protein folding using spin glass ideas. Then I will discuss the dynamics of these rather abstract models of the protein folding processes and show that, in some respects, we can regard the spin glass paradigm as a worst case analysis of the complexity of folding. Finally, in the last section, I will explain how ideas from the theory of associative memories can be used to construct an algorithm that may be useful in determining protein structure. This last effort is, of course, extremely preliminary and should be thought of as a glance forward to a possible resolution of the complete problem.

2. Phenomenological Background for the Protein Folding Problem[6]

A. Protein Architecture for Pedestrians

At the lowest level of description, a protein is a linear polymer whose backbone consists of a repeating unit — NHCHRCO. The stereo-typical nature of this repeating unit allows the same enzymatic machinery to assemble an arbitrary protein. The differences between proteins are determined by the sequence of the amino acids. The amino acids differ only in their side groups, R. This sequence of the amino acids is referred to as the "primary structure" of a protein. The genetic code allows the rewriting of a DNA sequence into a sequence of amino acid residues. Although under some conditions this linear polymer is random and uncoiled, like synthetic homopolymers in good solvents, the functioning folded protein is a collapsed structure with relatively well-defined three-dimensional structure. The three-dimensional structure is determined largely by the sequence of side chains, i.e., the primary structure of the protein. Folded structures of proteins with widely different sequences may have very much in common structurally. For example, the hemoglobin molecules of different animals and, indeed of plants, carry out very similar functions; namely, binding oxygen.[6] Nevertheless, their primary sequences differ in as much as 50% of the positions. It was only after the three-dimensional structure of hemoglobin was determined through the use of X-ray diffraction that it became apparent that the overall structure of the backbone in all of these different hemoglobins was very nearly the same. Even molecules which do not possess the same function can have very similar backbone structures. In fact, a taxonomic classification of protein structures is emerging.

At first sight, even the folded structure looks complicated and perhaps random. However, there are some rather important themes which allow one to pick apart the structure of a protein and make it a bit more comprehensible. First, when all the atoms are included in a structural representation of a protein molecule, one discovers that the molecule has very little empty space in it. It is rather different from the expanded polymers which one is used to dealing with in statistical mechanics. Rather, the atoms are nearly as closely packed as they are in amino acid crystals, although there are some cavities. It turns out that this close packing is achieved by exploiting some regular themes in the packing. A large part of the local structure of a protein is quite periodic. The protein must finally attain a compact shape, but in doing so, it divides into fairly rigid segments that have a repeating structure. The most well-known of these is the α-helix. In the α-helix, every fourth residue is hydrogen bonded to the first, creating a very rigid helical pattern. Another kind of local arrangement is the β-sheet in which the backbone takes on an extended form which presents groups capable of hydrogen bonding to another extended segment forming a β-sheet. These one-dimensional patterns could persist indefinitely in an infinite protein molecule. In fact, these themes are found in wool (the α-helix) or silk (the β-sheet) and essentially form the entire structure of such very long molecules.

In globular proteins, however, there must be other elements of structure that break up these infinite patterns and allow a compact structure to be formed. These are called turns. The assignment of these patterns, helix, sheet, or turn to different residues is called the secondary structure. For a long time it was thought that this secondary structure was simply encoded in the primary structure, and that certain patterns, locally, would define the secondary structure of any element. Indeed, there are strong tendencies for certain residues to take on different secondary structures, but no very simple rigid code of this sort has been found, although the search has been extensive and is continually being explored. The relatively weak stability of small segments of proteins suggests that such a rigid code is very unlikely to exist, but this is a controversial opinion. In forming a compact structure, these elements of secondary structure must themselves be packed together. This leads to the level of description called "super secondary structure". One can, for example, look at the way in which sheets may be packed onto each other. Just as in the problem of the packing of two-dimensional layers into close packed structures for crystals, there are only a few ways in which this may be done.[7] β-sheets, when placed on top of each other, must be oriented only in particular ways. Similarly, the α-helices can be packed on each other in only a few ways that allow close packing.[8] Coiled coils result which can be thought of as the packing of screws on top of each other. There are only certain paradigm angles which allow close packing of one helix on top of another. This observation was originally due to Francis Crick in the 1950's. It is clear that such close packings require some constraints on the side chains and, indeed, the side chains that are packed against each other typically are of very nearly the same steric volume, thus facilitating such regular packings. These "super secondary structure" correlations, however, still allow various three-dimensional arrangements, just as there are several ways of packing layers in close-packed three-dimensional solids leading to polytypes.

The final overall folding pattern is called the tertiary structure of a protein. The element of the side chains that seems to be greatly involved in taking up this tertiary structure is the requirement that the groups on the outside of the protein should be fairly soluble in water; that is, hydrophilic. Generally, but not always, the groups which are packing on the inside are hydrophobic; that is, the oil molecules. A very rough analogue to this pattern is the structure of a micelle in which the charged polar groups are on the outside and the oily groups are on the inside. The difference between a protein molecule and a typical micelle is that the internal arrangements correspond to these regular packings such as might be seen in a crystal.

At this writing, several hundred tertiary structures have been determined, using X-ray diffraction. One of the most exciting developments has been the realization that although there are many structures, they can be classified into between 20 and 30 families, corresponding roughly to different patterns of super secondary structure association.[9] Generally the large proteins are easier to classify in this way. At this point, the classification of families has been carried out using the biological insights of natural history, but it is clear to a physically-minded person that these pattern are analogous to the space groups, which one sees in three-dimensional crystals in some respects. At this point, only a small amount of work, albeit very insightful, has been done at *a prior* treatments of the possible families of tertiary structure.[10]

There are still higher levels of organization of proteins. Very large proteins, or even moderately large proteins, often can be subdivided into domains; regions which are contiguous in primary sequence and compact three-dimensionally.[11] Thus a domain would satisfy most of the architectural constraints we have discussed for the smaller proteins, but would bud off and have an extra domain connected to it. Often these domains can fold independently, and can be created as separate folded molecules in the test tube. There may be still smaller elements of modular organization as well, and this should be borne in mind when thinking about the folding problem.

Finally, functioning biomolecules are often really multiprotein complexes. The hemoglobin molecule, for example, is not a single covalently bound polymer, but is rather four molecules held together only by van der Waals interactions. This level of organization, the organization of protein molecules into protein complexes, is usually referred to as quaternary structure. Quaternary structure is often exploited in the function of biological molecules. The regulation of oxygen binding to hemoglobin is caused by modulation of quaternary structure.

This lightning review of protein architecture surely does not do justice to the field but, hopefully, for the novice, it dispels the notion that folded structures are random. Rather, they are very organized, albeit in a fairly complex way and in a way which we do not yet fully understand.

B. Thermodynamics and Kinetics of Folding and Unfolding[12]

A protein molecule is a finite system. Its transformations, therefore, can be thought of as chemical reactions. At the same time, a protein is a very large finite system in atomic terms. Thus one can also use the ideas of thermodynamics of phase transitions in order to discuss protein folding. It is important then to look at the experiments on folding from both the view point of chemical equilibrium and of phase equilibrium. Physical chemical studies of folding are usually carried out for systems in which unfolding can be induced by changing environmental conditions. Raising the temperature of a solution will usually cause unfolding. In some cases, a folded protein can be caused to unfold by lowering the temperature.[13] This is apparently because a large part of the binding forces involve the entropy of the surrounding solvent. In addition, folding can be induced by changing the pH of the solution or by adding compounds which change the water structure, such as urea or guanidium hydrochloride. At a fixed temperature or other environmental condition, one can study in dilute solution the relative concentrations of the folded and unfolded forms, thus obtaining an equilibrium constant which is a measure of the relative free energies of the two forms.

In order to determine the concentration of the forms, one must have probes which distinguish the states of the protein molecule. Some of these probes are physically quite direct and simple. For example, the viscosity of a protein solution is determined by the overall size of the molecules in it; thus, viscosity measurements can be used to probe whether the molecules are primarily compact or extended. Similarly, low angle X-ray scattering can be used to measure compactness. Some probes, such as visible spectroscopy or fluorescence, are sensitive to the local environment of certain residues in the protein, and therefore can be probes of secondary structure. In addition, the nuclear Overhauser effect and other NMR phenomena can give information about the relative location of two groups in a protein molecule and thus be a probe of tertiary structure.

In addition to these fairly direct physical probes, one can assay for folded and unfolded molecules with more biologically or biochemically relevant procedures. For example, one can raise antigens against the folded protein and use the binding of the antibodies to the protein solution to determine what fraction of the molecules exhibit antigenicity. This is a probe of local tertiary structure, or perhaps secondary structure. In addition one can do chemical crosslinking experiments in which a reagent containing bifunctional groups capable of forming crosslinks is added to the solution under one set of conditions. In order to see whether certain residues are near each other, the reagent is allowed irreversibly bind; then one can study at one's leisure the groups which were nearby during the crosslinking experiment. One can also use assays of the activity of an enzyme because the activity depends on the three-dimensional environment of the active site.

Finally, calorimetric studies can also be carried out which simply probe the total energy content of a protein solution, which can then be related back to equilibrium constants through Van't Hoff's law. All of these studies have been carried out on a huge variety of proteins. In this section I will summarize some of the general trends that one finds through these experiments, although the situation is often rather complex.

The first observation is that the free energy difference between folded and unfolded forms is often quite small, of the order of a few $k_B T$. This seems to be responsible for the complexity of even determining whether a protein will fold. As we have stated earlier, proteins with very different sequences can still fold into the same structure. On the other hand, a given protein can be modified even in fairly minor ways, leading to a molecule which does not undergo a folding transition under ordinary circumstances. The sensitivity to small sequence variations can be studied straightforwardly, using the tools of modern molecular biology to prepare proteins in which only a single site has mutated. This kind of study has become very popular of late. It is clear that even changing a single site can occasionally give rise to a situation in which no folding is observed. On the other hand, many site mutations change the free energy difference only in small amounts.

Similarly, one can study the folding of fragments of proteins to see whether this is possible. One of the classic studies that showed that folding can be reversible in many cases was Anfinsen's study of ribonuclease.[1] He showed also that if four amino acids were removed from the C-terminal end, the remaining fragment was unable to fold. On the other hand, in the case of the lysozyme molecule, the first twelve amino acids can be clipped off and the molecule will still fold. These observations alone indicate a very interesting degree of cooperativity in obtaining a structure. The interactions clearly involve forces between different parts of the molecule. This is most dramatically revealed in studies of complementarity. In the case of ribonuclease one can remove the first 20 amino acids and the remainder does not fold; however, if the first 20 amino acids as a peptide are added to the solution of the truncated protein, they will bind together and lead to a folded structure.

The static thermodynamic measurements also give more insight into the nature of the problem. Very often, one can see that as far as thermodynamics is concerned, the process is one involving two thermodynamically distinct states or phases. This can be seen through the fact that different probes (for example, viscosity probes or spectroscopic probes) indicate that the transition occurs at the same set of conditions as measured by each of the probes. When this is the case, one can also compare the calorimetry measurements directly with the spectroscopically measured equilibrium constants, and they agree. This, of course, does not mean that there are not any intermediate states in the folding but, rather, that in equilibrium their population is very small. In this sense we see that we have a situation very much like a first order phase transition in a finite system. There is an effective free energy with a double minimum structure, one minimum referring to states very much in the vicinity of the folded structure and the other minimum dealing with unfolded species. The intermediates must be higher in energy so that they are not present in large quantities. This, of course, means there is an energy barrier for folding or unfolding.

Two-state thermodynamics is not entirely universal, however. There are several examples in which different probes show non-coincident transitions. In this case, as well, there will be a lack of agreement of calorimetric measurements with a two-state equilibrium. There seem to be at least two origins of these non-coincident transitions. In the case of multiple domain proteins, it seems that one of the domains can unfold prior to the other one. Another important source of non-coincidence in single domain proteins is that one can obtain a compact structure which is only slightly larger than the folded protein in which there is very little native secondary and tertiary structure.[14] This compact structure generally fluctuates more than the folded native structure and is sometimes referred to as the molten globule state.

In addition to equilibrium measurements, kinetics can be very revealing. In general, unfolding is a simple exponential process. One reaches a rate limiting transition state which destabilizes the folded molecule and the further steps are rather rapid. In folding, on the other hand, the behavior is often typically biphasic, characterized by two exponentials, one with a short time constant and a longer one. Again, this is not to say that the process is simply characterized by two intermediates. In fact, generally, as more probes are introduced one finds more intermediate states, although many of these exhibit very similar energies and have facile transitions between them. This study of intermediates has only really begun although for some simple proteins, which are easily crosslinked, many intermediates have been discovered through trapping experiments as long ago as the 70's.[15] As NMR technology improves, one can anticipate more studies of this sort.

In addition to the relatively "clean" observations that I have summarized above, there are often some glaring examples of more perplexing behavior implicating much longer timescale dynamics. It is quite common outside of the small protein realm to find that the transition for unfolded to folded states is imperfectly reversible.[16] For example, in the case of elastase, the optical signal of folding is perfectly reversible. On the other hand, the activity is not completely reversible after unfolding and the antigenic properties are not completely recovered. Clearly these biological measures of structure are more sensitive than the simple visible spectroscopic probe. Many other examples of irreversibility exist. The usual explanation for these sorts of situations is that there is competition with aggregation; that is, a partially folded or misfolded molecule can aggregate with another molecule to form a larger complex and, perhaps, lead to large colloidal particles. Often, however, this aggregation does not completely account for the loss of reversibility. In any event, this phenomena would mean that there are very long-lived misfolded intermediates. In addition, there have been very recent observations of catalysis of folding and assembly,[12] which again, implicate a much more complicated spectrum of lifetimes of intermediates. It is in this problem of misfolding where the theory of spin glasses can help in our understanding since it is clear that we must account, in these cases, for a diversity of states.

3. Spin Glasses and Protein Folding

A. A Statistical Protein Model

As we have seen, in an overall sense, the process of folding is much like a first order phase transition in a finite system. The finally folded protein, although it is complex, has themes reminiscent of extended crystals. However, it is clear that in the thermodynamic and dynamic experiments, one can see revealed intermediate states during folding. Some of these may correspond to collapsed structures and, in addition, there may be a plenitude of differently folded structures. The usual theories of first order phase transitions do not take into account these intermediates because the free energy functions are imagined to be smooth functions of the relevant order parameters. One way to get a handle on these other states might be to introduce a statistical model of the energies of intermediately folded proteins.[17] As we shall see, in some sense, this is a worst case analysis of the kinetic difficulty of folding. In any event, the analysis is simple and suggests new possibilities for the dynamics of folding and new constraints on computer models of folding.

Before starting this description of a statistical protein model, it is perhaps useful to look at a naive mean field theory of the interplay of folding and collapse. This will be along the lines of the Flory theory of excluded volume.[18,19] In this mean field model, one can use two overall order parameters to characterize the system. One of these related to the collapse is a measure of the size of the partially folded protein molecule, a radius of gyration, say R. In a coil state, R will be large while in a folded or molten globule state, the radius will be relatively small. In addition, then, one requires an order parameter measuring how close the system is to the native structure. A simple measure of this is the fraction of amino acid residues whose conformational angles are within some specified amount of the angles they retain in the completely folded structure. This we will call the fraction of native residues n_0. By using this choice we have taken a stance that secondary structure is rather important. Other possible measures of similarity to the native may be used; for example, whether appropriate contacts are made for segments distant along the chain. The energies of states consistent with a given value of these order parameters will, of course, vary, but it is possible, since we know that a folded structure eventually results, to make an estimate of the average energy of states with given values of the order parameters. The energies associated with the secondary structure will simply be proportional to n_0^2 because one is concerned merely with whether nearby segments in sequence are in their correction stereochemical configuration; thus, there is a contribution to $\bar{E}$ given by $\bar{E} = -An_0^2$. If there were only these secondary structure considerations, one would have a one-dimensional problem and the mean field approximation would be exceedingly poor. In fact, in polymers that do not collapse, the secondary structure transition that results is essentially like that of the one-dimensional Ising model and a large literature on such helix coil transitions, for example, exists.[20,21]

On the other hand, in a collapsed structure, segments which are distant in sequence will come together, and if these are in their correct stereochemical orientations, a stabilizing interaction results as we discussed in the theory of packing considerations. Thus there is a non-local contribution to the average energy $\bar{E} = -B(R)n^2$. The coefficient, $B(R)$, is proportional to the probability that two segments will be next to each other in a coil or globule of radius R. Thus it greatly increases as R decreases. The sum of these energy terms greatly encourages the formation of collapsed and correctly folded structures.

The coil structure, on the other hand, is favored by entropic considerations. Each of the incorrectly folded non-native amino acid residues contributes to the entropy of the coil. With no constraint on the size parameter then, one obtains a number of states, $\Omega_0 = \nu^{N(N-N_0)}$ where

$N_0 = Nn_0 \cdot \nu$, the conformational freedom of non-native segments is of order 10 from various estimates. The collapse order parameter R also affects the entropy for small R. The number of configurations confined to a region between R and $R + dR$ is given by:

$$\Omega(R) \equiv \nu^N \left(\frac{C}{R_0}\right)\left(\frac{R}{R_0}\right)^2 e^{-\frac{3}{2}\left(\frac{R}{R_0}\right)^2} \tag{1}$$

R_0 is the mean radius of gyration associated with the "random" bonds. Thus, $R_0 \sim (N - N_0)^{\frac{1}{2}}$. The entropy of the set of configurations with order parameters n_0 and R is given by the logarithm of this number

$$S^* = k_B \ln \Omega. \tag{2}$$

Combining the entropy and the mean energy, thus ascribing the same energy to each of the micro states, one obtains a free energy as a function of the order parameters given by:

$$F(n_0, R) = -k_B T S^*(n_0, R) - [A + B(R)]n_0^2. \tag{3}$$

This function can have essentially two minima corresponding to: (1) an unfolded structure with very little native structure with high entropy and high energy; (2) a native structure which is collapsed and in which the energy is quite low and the entropy quite low, and R only modest in size. One ends up with a typical 2 minimum potential with a barrier in between that must be overcome by a nucleation event. In this mean field description, the nucleation event must involve a large fraction of the entire molecule.

The description of the statistical protein model[17,22] adds only one feature to the analysis we have made, à la Flory. Instead of ascribing to each configuration consistent with an overall value of the order parameters a single energy, one imagines that each configuration has a random energy. Furthermore, one asserts that the energies of different conformational states consistent with the same value of the order parameter are uncorrelated. This is certainly an oversimplification, although one can imagine that because conformational changes are bringing wildly different parts of the protein in contact, it may not be as strange for proteins as it may seem for other systems. Once this assumption is made, one needs to characterize the probability distribution of these energies. The mean of this probability distribution is given by the same expression as before, but one must also specify the variance. The variance arises from bringing together improperly folded parts. Thus we expect the variance to be larger in collapsed states because there are more contacts and larger when there is a great deal of a non-native structure. A rough form for the variance then, is

$$\Delta E^2 = C \, B(R)(1 - n_0)^2 \,. \tag{4}$$

Our ansatz is essentially the ansatz that Derrida has made in his random energy model for spin system,[23,24] simply transcribing it to the polymeric context. It is easy, then, to follow his analysis with the extra introduction of the order parameters, R and n_0. The easiest route to the thermodynamics is through the entropy. At a given value of the order parameters, the probability distribution of energies of states is a Gaussian:

$$P(E; N_0 R) = \frac{1}{\sqrt{2\pi\Delta E(n_0, R)}} \, e^{-\frac{(E - \bar{E}(n_0, R))^2}{2\Delta E^2}} \,. \tag{5}$$

Thus, the average number of states of the well-defined energy is given by $\Omega(E) = \Omega^* P(E)$. This gives the entropy as a function of energy E, and one can solve for the corresponding temperature using the thermodynamical relationship $1/T = \delta S/\delta E$. One can now, in a more familiar manner, rewrite the free energy as a function of the order parameters and entropy.

$$F = -Ts^*(R, n_0) - \bar{E}(R, n_0) - \frac{\Delta E^2}{2T} \, (R, n_0) \,, \tag{6}$$

$$S = S^* - \frac{\Delta E^2}{2T} \, (R, n_0) \,. \tag{7}$$

Notice that the equation for the free energy is of essentially the same structure as the one described for the Flory mean field theory of folding and collapse. The only difference is that the energy term contains the variance as well as the mean energy. This arises because at thermal equilibrium, one will tend to probe the lower energy states of the distribution rather than the mean, although this will be a balance determined by the loss of probability at energies much lower than the mean. At the level of the free energy, the statistical protein model has much the same behavior as the mean field protein model. There are two new features. First the interaction can now be favorable for collapse even if no native structure is formed because of the variance term. This can lead to a new state: the molten globule. Second, the entropy has a peculiar behavior. Again, because of the tendency to occupy lower energy states, the entropy decreases with temperature, even at a fixed value of the order parameters. At the temperature T_g, the entropy which is entirely configurational would vanish. What happens below this temperature? Clearly the entropy does not go negative; rather, we see that the approximation of the probability distribution by a continuous one smooths out, as a Gaussian must be in error in this regime. The real distribution is a histogram, a set of spikes. Once the entropy goes to zero, this means one is probing the lowest tail of this distribution. Only a few states are present there, and the smoothed approximation is a bad one. In this regime one would better write that there is a single state whose energy is determined by the condition $S(E) = 0$ and the thermodynamics is frozen at this point. In fact, the study of the random energy model shows that there is not merely a single state, but rather there is but a thermodynamically insignificant number. For our purposes, this distinction is not essential.

The entropy crisis then indicates that in collapsed (molten globule) or partially folded states one can have at a sufficiently low temperature or sufficiently random interactions a glass phase which consists of very stable misfolded structures. In the folded protein, because there will be so few non-native residues, this glass phase will occur at a fairly low temperature. That transition is perhaps related to the glass transitions seen in folded protein as discussed in the article by Austin in this volume. (Although, it is not at all clear that the approximations that we have made so far are relevant for such experiments.) Alternate and more likely views of the glass transitions seen in folded proteins are that they are due to interactions of conformational degrees of freedom of the sidechains[24] or the solvent.[25,26] The molten globule state, on the other hand, has a great deal of randomness and is, perhaps, more susceptible to forming a glass even at higher temperatures. Such misfolded states would represent a dead end along a folding pathway from which it would be impossible to get to the folded structure.

More important than the possible existence of equilibrium glass phases is the role that the proximity of a glass phase may play in the dynamics of folding. As the folded structure is approached but not yet reached, one can have an intermediate value of the order parameters where the glass transition is close. This will have an effect on the rate of nucleation and may be involved in the difficulty of folding some proteins. For this we must investigate the dynamics of the model.

B. Dynamics of the Statistical Protein Model

The dynamics of the statistical protein model can be studied approximately.[27] The way in which one can begin is by first getting some idea of the statistics of the minima of the energy function. The key questions are: How many minima are there? How deep are the basins for these minima? What is a typical escape time distribution? The consideration of these problems is very similar to issues involved in the diffusion of particles in disordered media. However, the high dimensionality of the problem allows both simplifications and complications.

To make this analysis of the statistical topography, we must have a relevant definition of the connectivity of the microstates of the protein. We take a very simple dynamical scheme in which a protein changes its state by allowing a single amino acid to take on a new conformation. Over the years a great deal of work has been done in polymer physics about realistic models of conformational changes in polymers and, in general, this sort of single conformational motion is found to be much more probable than a concerted one.[28] Thus, if there are N amino acids in the protein, each microstate has $N\nu$ neighbors. The total number of single flip minima is quite easy to estimate. Consider a given state and its $N\nu$ neighbors. Within the random energy model, these $N\nu+1$ states are statistically independent. Thus, there is equal possibility that any given one of them is the lowest. Therefore, the probability that the central state is the lowest is $1/(N\nu + 1)$. Thus the number of single flip local minima[29] is enormous.

$$N_m = \frac{\nu^N}{N\nu + 1}.\tag{8}$$

The energy distribution of these local minima requires a bit more analysis. Obviously, a high energy state is going to have a low probability of being a minimum. On the other hand, there is an enhanced probability of a low energy state being a minimum since its neighbors are more likely to be near the mean energy. However, very low energy states themselves are improbable. Thus, more minima will be somewhat below the average energy, but should not be very much below this energy, typically. A more precise estimate can be made in the following way. We see that the distribution is the product of the probability that a given minimum of energy, E_0, exists, times the probability that all of its neighbors have higher energies. Let $g(e, N_0, R)$ be the probability distribution for the energies at a given value of the order parameter. The probability that all $N\nu$ neighbors are higher in energy than E_0 is given by

$$P_{M,N}(E_0, n_0, R) = \left[\int_{E_0}^{\infty} dE g(E, n_0, R) \right]^{N\nu}. \tag{9}$$

Since $N\nu$ is large, this can be written in another way

$$P_{M,N}(E_0, n_0, R) = \exp\left[-N\nu \int_{-\infty}^{E_0} g(E, n_0, R) de \right]. \tag{10}$$

In the limit of large $N\nu$ the distribution of minima becomes a truncated Gaussian

$$\begin{aligned}
g_{M,N}(-E_0, n_0, R) &= g(E_0, n_0, R) P_{M,N}(E_0, n_0, R) \\
&= \frac{N\nu}{\sqrt{2\pi\Delta E^2}} \exp -\left(\frac{(E_0 - \bar{E})^2}{2\Delta E^2} \right) \quad \text{for } E < E_c \\
&= 0 \quad \text{for } E > E_c
\end{aligned} \tag{11}$$

where the truncation point is an energy somewhat lower than the mean energy,

$$E_c = \bar{E} - \sqrt{2 \log N\nu} \, \Delta E. \tag{12}$$

The escape time, $t = k^{-1}$ from a local minimum is somewhat more difficult to estimate. In addition to a notion of locality, we need to apply a rule for generating dynamics from the energy surface. The simplest scheme to use, which we shall employ, is that of Metropolis dynamics. The rate of going from a high energy state a to a lower state b is k_0; that is, all downward moves are accepted at a rate k_0; all upward moves are accepted only occasionally when an activation energy has been accumulated, so that detailed balance is satisfied:

$$k_{ba} = k_0 \, e^{-(E_b - E_a)/k_B T}.$$

We will make the approximation that a given protein molecule has escaped from a local minimum when it has successfully made a single conformational change. This would seem to be a good approximation for the random energy model with large $N\nu$ because the near neighbors are themselves rather unlikely to be also minima with respect to further moves. The rate of leaving a local minimum, then, can be given by

$$k = k_0 \sum \exp -\{E_i - E_0\}/k_B T \qquad (13)$$

where E_i is the energy of each connected state. Since the connected states have random energies consistent with being larger than E_0, their distribution is given by

$$g_{con}(E_0, n_0) = \frac{g(E_0, n_0)}{\displaystyle\int_{E_0}^{\infty} g(E, n_0)dE} \quad \text{for } E > E_0, \quad 0 \text{ otherwise}. \qquad (14)$$

The quantity k will itself be distributed, but its average value $\bar{k}$ is easy to compute. Knowing g_{con}, one obtains

$$\bar{k} = k_0 N\nu \exp\left[\frac{E_0 - \bar{E}}{k_B T} + \frac{\Delta E^2}{2(k_B T)^2}\right.$$

$$\left. \times \left\{\frac{1 - \text{sign}\left(E_0 - \bar{E} - \frac{\Delta E^2}{k_B T}\right)\text{erf}\left(E_0 - \bar{E} + \frac{\Delta E^2}{k_B T}/2\Delta E\right)}{1 + \text{erf}\left(\frac{E_0 - \bar{E}}{2\Delta E}\right)}\right\}\right]. \qquad (15)$$

For large proteins, this function has two behaviors. For low energy states, $E_0 < \bar{E} - \Delta E^2/2T$ which is, in fact, the thermally weighted average energy over the random energy state; $\bar{k}$ is given by

$$\bar{k} = k_0 N\nu \exp\left[\frac{E_0 - \bar{E}}{k_B T} + \frac{\Delta E^2}{2(k_B T)^2}\right]. \qquad (16)$$

Notice that this equation essentially corresponds to saying that there is a positive entropy of activation $k_B \log N\nu$ corresponding to an escape path to any of the neighbors, and an energy of activation essentially equal to the difference in starting energy and the thermal mean energy of connected states. For minima more energetic than the thermally averaged energy, the average rate is given by

$$\bar{k} \equiv k_0 N\nu \exp\left\{-\frac{(E_0 - \bar{E})^2}{2\Delta E^2}\right\}. \qquad (17)$$

If we ignore the fluctuations in escape rate from a site of given energy, we see that the distribution of escape will essentially mirror the distribution of minima. The deep minima are essentially Gaussian distributed and their activation energy is that to reach the mean; thus the probability distribution for the slower rate is given by a log normal distribution

$$P(k) = \frac{1}{\sqrt{2\pi}k_0} \frac{T}{\Delta e} \exp - \left\{ \frac{(k_B T)^2}{2\Delta E^2} \left[\log\left(\frac{k}{k^*}\right) \right]^2 \right\}. \qquad (18)$$

The faster rates, those with $k > k^*$, have a different distribution

$$P(k) = \frac{1}{\sqrt{2\pi}} k_0 \frac{1}{\sqrt{2\log\left(\frac{k_0 N\nu}{k}\right)}}. \qquad (19)$$

The separation rate, k^*, is given by

$$k^* = k_0 N\nu \exp - [\Delta E^2 / 2k_B^2 T^2]. \qquad (20)$$

k^* is in some sense a typical rate of escape.[30] It is useful to write this typical rate of escape in terms of the configurational entropy, s^*, and the glass transition temperature:

$$k^* = k_0 N\nu \exp\left\{ -\frac{S^*}{k_B} \left(\frac{T_g}{T}\right)^2 \right\}, \qquad (21)$$

This typical escape time gets longer as the glass transition temperature is approached until at T_g itself, the typical escape time is

$$t^* = \frac{1}{k_0 N\nu} \exp \frac{S^*}{k_B}, \qquad (22)$$

This is quite a picturesque result. Notice that it indicates that there is effectively only one other state to which one can escape; that is, there is a huge negative entropy of activation relative to the expected result. The typical minimum at this temperature will be one of the ground states. Therefore, the only possible escape will be to another one of the quasi-degenerate ground states, which are not thermodynamically significant in number. These local aspects of the statistical topography must be combined with a picture of the overall folding process to estimate the time scale of folding. Once we have a distribution of lifetimes, the simplest model to use is the continuous random walk, due to Montroll and Shlesinger.[31] We have used this to examine the folding process and to get some idea of the role of distribution of times in the process.[27]. However, a very straightforward derivation of the results may be obtained by assuming that the overall process involves a diffusion in order parameter space, N_0:

$$\frac{\partial P(n_0)}{\partial t} = \frac{\partial}{\partial n_0}\left[D(n_0)\frac{\partial P}{\partial n_0} + P\frac{\partial F(n_0)/k_B T}{\partial n_0}\right]. \qquad (23)$$

The free energy, as a function of n_0, is given as a result of our developments in the last section as a double minimum potential. The diffusion constant is essentially the expectation value of the rate leaving the minimum energy site. The resulting diffusion constant has a complex form, but it basically can be well approximated in two regimes for $T > 2T_g$, it is

$$D(n_0) = \frac{1}{2}\frac{k_0}{N}\exp-\left\{\frac{\Delta E^2(n_0)}{k_B^2 T^2}\right\} \qquad (24)$$

and for T between $2T_g$ and T_g,

$$D(n_0) = \frac{1}{2}\frac{k_0}{N}\exp\left\{-S^*(n_0) + \left(\frac{1}{k_B T_g} - \frac{1}{k_B T}\right)^2\Delta E^2(n_0)\right\}. \qquad (25)$$

Notice that these two forms are continuous at the break point and only differ slightly if the first one were extrapolated to T_g.

In terms of this order parameter, the problem is one of single dimensional diffusion in a double minimum potential. The diffusion equation can be integrated[33,34] to give a mean first passage time:

$$\bar{t} = \int' dn_0' \int^{n_0'} dn_0 D(n_0)^{-1}\exp\left\{\frac{1}{k_B T}\{F(n_0) - F(0)\}\right\}. \qquad (26)$$

Because T_g depends on the extent of folding as well as S^*, D in this expression can be a strong function of the order parameter, n_0.

It is clear in examining this that there are two possible limits: One is that in which the diffusion constant, even in the unfolded state, is quite small, such as may occur if the unfolded minimum is collapsed. On the other hand, it may happen that the most rapidly varying term is the activation free energy itself, in which case the diffusion constant at that activation free energy is all that is required. This leads to a modified Kramers expression where the diffusion constant is evaluated near the transition state. In this instance what is relevant is whether a glass transition occurs near the transition state for folding, not whether the collapsed starting state is glassy. In general, the configurational entropy will be smaller at this transition state, thus the overall diminution in rate for the search problem in this vicinity is certainly smaller than it would be if there were a collapsed state. If a glass transition occurs along the path to folding, then there will be a statistical bottleneck. The rate of folding will be essentially that of exploring all configurations in this vicinity; that is, one has achieved the horrendous limit in which search is most difficult, rather than simply having an activation barrier to overcome.

This nightmare was anticipated long ago by Levinthal in his treatment of the question of pathways to folding.[34] Here we see that it is not a problem unless glassy states emerge on the pathway. Ironically, the existence of a glass transition would be equivalent to having a unique series of states through which one must pass and is thus the limit of a very *unique* pathway. Avoiding a glass transition is equivalent to having many ways in which folding can occur and having a relatively non-unique pathway up to a nucleation transition state.[35] Apparently, then, easily folded proteins do not encounter a glass transition along their folding pathway.

An equally important conclusion is that in designing models for simulating protein folding, and perhaps predicting protein tertiary structure, one should investigate models with low nucleation barriers and with very low glass transition temperatures. These criteria, we believe, will be helpful in designing protein folding Hamiltonians.

4. Sophisticated Theories of Heteropolymer Collapse as a Basis for the Statistical Protein Model

The statistical protein model is easy to analyze. However, it is not clear what is the class of realistic Hamiltonians which would be described effectively by this model. A variety of models of heterogeneous polymers have been investigated using techniques of spin glass theory.[36-40] In general these models suggest a reasonable faithfulness to the statistical protein picture. In this section, I will review the techniques used and results obtained by these various authors.

Many of these more sophisticated theories have begun by the consideration of a problem which is better considered to be random heteropolymer collapse than folding; that is, the interactions are taken to be random functions of the sequence without a special concern that there be, *a priori*, a single well-folded structure. Furthermore, to use the apparatus of polymer theory, the interactions are generally taken to be of contact form, $V_{ij} = V_{ij}\delta(r_i - r_j)$. The partition function, then, takes the form of

$$Z = \int dr_1 \cdots dr_N \prod_{i=1}^{N-1} \delta((r_{i-1} - r_i) - a) \prod_{i<j} e^{-\beta V_{ij}(r_{ij})} \qquad (27)$$

where $\beta = 1/k_B T$, sets the temperature scale. One then computes the quenched free energy, assuming that the V_{ij} are independent Gaussian variables with a distribution given by

$$P(V_{ij}) = \prod_{i<j} \frac{1}{\sqrt{2\pi \cdot V^2}} \, e^{-\frac{1}{2V^2} V_{ij}^2}. \qquad (28)$$

A hard core repulsion may also be added in several of the treatments.

Using this form of the interaction partition function, Garel and Orland[36] have argued that when the sites are confined to a lattice, as opposed to being in the continuum, then the chain constraint is of little relevance and the system has a transition like that of a Potts spin glass. The Potts spin glass has been shown to have a phase transition of essentially the same type as the random energy model involving an entropy crisis.[23,41] Hence, according to the Garel/Orland analogy, the statistical protein model would seem reasonable. Garel and Orland were, however, concerned that the chain constraint would play an important role, as we have found with the simple Flory theory as well. Their argument was based on the result that the scaling exponent for the Potts glass is equal to $2/d$, which differs from the result $\nu = 1/2$ for a random walk, which would be expected for an extended chain. This concern seems to be based on the idea that the collapse transition to a globule and the glass transition must be coincident, which the statistical Flory theory indicates is not the case.

In any event, these considerations led Garel and Orland to examine the partition function in the continuum limit. Here the partition function for a single chain can be written as a path integral:

$$Z = \int Dr(s) \exp - \int ds \frac{\dot{r}^2}{2} - \beta \int ds \int ds' v_{s,s} \delta(r(s) - r(s')). \qquad (29)$$

This path integral expression can then be used to find the quenched average of the free energy, using the replica trick:

$$\langle \log Z \rangle_v = \underset{n \to 0}{\mathrm{Lt}} \frac{\langle Z^n - 1 \rangle}{n}$$

$$\langle Z^n \rangle = \int Dm_\alpha(r) \prod_{\alpha < \beta} Dq_{\alpha\beta}(r, r) \, \zeta$$

$$\times \int \exp - \left[\beta^2 v^2 \sum_\alpha \int dr \, m_\alpha^2(r) - \frac{\beta^2 v^2}{2} \iint dr \, dr' q_{\alpha\beta}^2(r, r') \right],$$

where

$$\zeta = \int \prod_\alpha Dr_\alpha(s) \exp - \left[\sum_\alpha \dot{r}_\alpha^2 + \beta^2 v^2 a^{-1/2} \sum_\alpha \int ds \, m_\alpha(r(s)) \right.$$

$$\left. + \beta^2 v^2 a^{-1} \sum_{\alpha\beta} \int ds' \, q_{\alpha\beta}(r(s), r(s')) \right] \tag{30}$$

ζ is the partition function of n replica polymers in a field m_α and with interactions between replicas $q_{\alpha\beta}$. As usual, the auxiliary fields in the replica problem represent fluctuations in the density of the polymer and the correlation of polymers in different replicas.

Notice that the Edwards/Anderson order parameter, $q_{\alpha\beta}$, is a measure of how related different polymer states are, and is somewhat like the frozen disorder contribution to the Debye-Waller factor, which is used to characterize X-ray diffraction studies of proteins.[42] If the saddle point approximation is used for the m and q fields, we see that the problem reduces to the determination of the partition function of a set of n polymers in a single self-consistently determined mean field and with a single-consistent interaction between replicas:

$$M_\alpha(r) = \frac{1}{2} \sum_i \langle \delta(r - r_i) \rangle \tag{31}$$

$$q_{\alpha\beta}(r, r^1) = \sum \langle \delta(r - r_i^\alpha) \delta(r - r_i^\beta) \rangle . \tag{32}$$

In the first treatment, due to Garel and Orland, one solves the interacting polymer problem by assuming that the minimum solutions are replica symmetric.

This replica symmetric ansatz leads to a collapse transition which differs little from that in homopolymers. To obtain realistic results, one should include the effects of excluded volume in the original functional integrals. Unfortunately the replica symmetric ansatz for the heteropolymer problem forces the collapse and freezing to occur simultaneously, thus the connection to the Potts glass is not captured by these approximations.

Shakhnovich and Gutin have treated the quenched polymer problem more completely in order to distinguish the collapse phase from the frozen phase.[38–40]. They point out that it is useful to carry out the analysis in two stages. They argue that the replica symmetric field m_α leads to collapse to a globule with characteristics of standard homopolymer collapse.[19] The m field then confines the polymer to a drop with a mean density $\bar{\rho}$, given by a balance of excluded volume and attraction. The surface terms in $m(r)$ are then neglected (although this is probably not essential).

The glassy phenomena involve the interaction between different replicas. In their theory then, the overlap parameter, $q_{\alpha\beta}$, is not simply related to the mean density as it is in the replica symmetric theory. The concomitant ordering field will localize particles to regions, which can be much smaller than the globule size. In the Shakhnovich–Gutin theory, with these approximations, the polymer chains can be eliminated to give a free energy function of the overlap order parameter, $q_{\alpha\beta}$:

$$F = E(q_{\alpha\beta}) - TS(q_{\alpha\beta}),$$

where

$$E(q_{\alpha\beta}) = \frac{v^2}{2} \int \sum_{\alpha,\beta} q_{\alpha\beta}^2(r, r')dr\ dr',$$

and

$$S(q_{\alpha\beta}) = k_B \log \int Dr_\alpha(s) \exp - \left[\sum_\alpha \dot{r}_\alpha^2 + \sum_{\alpha,\beta} q_{\alpha,\beta}(r_\alpha(s), r(s')) \right]$$
$$\times \prod_{\alpha<\beta} \delta[\delta(x_\alpha(s) - r)\delta(x_\beta - r') - q_{\alpha\beta}(r, r')]. \tag{33}$$

The evaluation of this free energy function still has many points in common with the theory of homopolymer collapse (but with several important differences).

The entropy of the replica polymer chains with a defined value of the Edwards/Anderson order parameter can be computed using the path integral. Alternatively, since the polymers are long, one can assume that there is ground state dominance.[43] The entropy of the n polymer chains is then the kinetic energy of a ground state wave function, $\Phi(\{r_{ij} - r_N\})$:

$$S = \frac{1}{2} a^2 \int \sum_\alpha (\nabla_\alpha \Phi)^2 \prod_\alpha dr_\alpha . \tag{34}$$

In terms of this ground state wave function the Edwards–Anderson order parameter is $q_{\alpha\beta}(r, r') = \int d\{r\}\phi^2(\{r\})\delta(r_1 - r)\delta(r_2 - r')$; thus, we obtain an equation for the free energy as a function of this ground state wave function:

$$F(\Phi) = -k_B T S(\Phi) - \frac{1}{2} B^2 \sum_{\alpha=\beta} \int d\{r\}d\{r'\}\Phi^2(r)\Phi^2(r')$$
$$\times \delta(r_\alpha - r')g(r_\beta - r') . \tag{35}$$

The variational equation for ϕ also results

$$(a^2 \Delta^2 + M)\Phi = -2B^2 \Phi(R) \int \sum_{\alpha\neq\beta} \Phi^2(V^1)\delta(r_\alpha - r^1)\delta(r_\beta - r)dr^1 , \tag{36}$$

μ is a Lagrange multiplier that guarantees that $\int \Phi^2(r)d\{r\} = N$, corresponding to confinement to the globule. The scale of these wave functions determines the mean size of fluctuations around a given structure which can be found. If this scale is denoted R_0, one can show that:

$$F(R_0) = \frac{A_1}{R_0^2} - \frac{A_2}{R_0^d} . \tag{37}$$

Notice that if the dimension d is greater than 2, only two values of this scale are possible. Either $R_0 = \infty$, corresponding to no freezing, or $R_0 = 0$; that is, R_0 is of the order of microscopic scales not included in the continuum description of the free energy. This sort of jump of R in a system with disorder is typical of transitions of the Potts glass type. Shakhnovich and Gutin, using a more microscopic description,[39] have, in fact, shown that this is of the Potts variety for $d > 2$.

Shakhnovich and Gutin also analyze the situation for $d < 2$. In this case, a continuous transition occurs that can be analyzed thoroughly using replica methods. In the low dimensional case, there is a series of replica symmetry-breaking transitions leading to a continuous Parisi order parameter, $q(x)$. In the polymer context this would seem to imply the existence of a clustered frozen polymer. Such a state would be very interesting for investigation. Its relationship to three-dimensional proteins, of course, is still unclear.

Another kind of heteropolymer system has been investigated by Garel and Orland.[37] Instead of taking the interactions V_{ij} to be Gaussian random variables, one takes them to be a sum of separable interactions

$$V_{ig} = -\frac{1}{N} \sum_{P=1}^{M} v_P \xi_i^P \xi_i^P \tag{38}$$

where the η_i^P are independent random variables. The interactions are still taken to have the contact form. Garel and Orland show, using similar reasoning as the random case, that a transition can occur for these models. This transition is of the Mattis type for a ferromagnet and essentially corresponds to collapse to a state with specific interactions between the subunits. This Mattis type Hamiltonian has some connection with the associative memory Hamiltonian to be discussed in the next section. One would expect, in addition to the transition elucidated by Garel and Orland, that if a sufficiently large number of characteristics are included, a spin glass-like transition would occur.

All of the heteropolymer systems have fixed sequences and, therefore, the interactions are fixed, but there are no explicitly frozen constraints between monomers on the chain. In some proteins, specific monomers can be crosslinked, e.g., the cystines by the formation of covalent disulfide bonds which are sufficiently stable that they do not thermally equilibrate in the usual physiological environment. This sort of crosslinking is very analogous to the vulcanization of rubber. Goldbart and Goldenfeld have used spin glass techniques to examine the vulcanization transition in uniform polymeric gels.[44] They conclude that there is a critical number of crosslinks which are necessary to give a frozen structure. Their analysis also gives a spin glass transition of the Ising sort. It is, however, possible that higher order terms convert this transition to one of the Potts variety.[45]

Although chemical crosslinking is a special mechanism giving stability to proteins, there is a more general type of crosslinking which one needs to investigate to understand the computational problem of determining tertiary structure. Nuclear magnetic resonance experiments allow one to determine bounds on the distance between residues in proteins. These bounds can be used as information in determining protein structures.[46] Such constraints on distances can be incorporated into a thermodynamics formalism by including crosslinks with these desired bounds in a Hamiltonian. Thus, when one is investigating the computational difficulty of folding, using as input experimental NMR distance information, the same heteropolymer theories of vulcanization will likely prove useful.

5. Using Spin Glass Ideas in the Design of Protein Folding Algorithms

The spin glass perspective provides an interesting way of looking at the degree of difficulty that the protein folding problem possesses. Can we go beyond the worst case analyses that we have just described in order to find new ways of inferring protein structure from sequence? Answering this question will take a great deal of work. The first step is to obtain models that contain the features of the statistical protein model, but which have realistic elements of protein structure in them. Models that describe only roughly the nature of phase transitions in disordered polymers will be enlightening but insufficient. An attempt to develop models that have realistic aspects of protein structure in them has recently been made by Friedrichs and Wolynes.[47] The attitude they have taken is that the development of simplified models for folding is equivalent to a problem in inverse statistical mechanics. That is, while in statistical mechanics one tries to find the consequences for structure and dynamics of a given Hamiltonian, in protein folding one is being asked instead to find a simple Hamiltonian which will give known structures. If such a simple Hamiltonian is found and its relationship to the sequence is understood, then one can hope to design a Hamiltonian which will allow folding to take place for a protein with a new sequence. This inverse statistical mechanics problem is, in a sense, much like the problem of statistical inference or pattern recognition, using many examples to find a law which will allow you to predict new instances.

From this way of looking at things, then, the design of a Hamiltonian for protein folding has much in common with the problems of associative memory and connectionist neural modeling,[3] which have been discussed elsewhere in this volume. This first step, then, is to find the polymeric analog of the associative memory spin models introduced by Hopfield. A polymer associative memory Hamiltonian can be obtained from the Hopfield associative memories by using the famous lattice gas analogy between fluid and magnetic systems. Density fluctuations play the role of spin coordinates in that analogy. For proteins, we must take into account that the various amino acids differ from each other. These differences can be encoded in various "charges". These "charges" are measures of hydrophobicity, size or electrical charge that might be ascribed to the amino acid residues. The tertiary structure to be associated with sequence then involves various arrangements of these charges in space. The center of charge can be taken as residing at the coordinates of the α-carbons of the protein. To make the analogy precise, imagine embedding the protein in a four-dimensional grid. Three of these dimensions are the space dimensions, and one of them is the sequence number, playing the role of time in the path integral treatments described in the last section. Any given cell in this grid will either be empty or occupied. Each occupied cell will have a charge density with which we can label that cell. The charge density of ± 1 can be assigned to cells which are occupied. We found in calculations that the hydrophobic charge density was the most convenient to use. On the other hand, empty cells will have a spin of zero. In this four-dimensional representation, proteins are a very sparse filling of the cells. Ideas from spin glass theory for sparse memories may be useful here.[48]

In this manner, a protein is represented as a spin pattern on the four-dimensional lattice. Once a selection of data base proteins has been made for the memories, one can now write down a Hamiltonian that has the corresponding spin patterns as minima, using the usual prescription of Hopfield:

$$H = -\sum_{ij} \sum_{\alpha} \lambda \, S_i^\alpha S_j^\alpha S_i S_j . \tag{39}$$

The potential energy of interaction between the sites is the correlation function of the charge densities with respect to space and sequence in the original data base. (Notice this is like the "inverse random phase approximation".) In the lattice gas analogy, one can then minimize this Hamiltonian, allowing only configurations that can represent an actual protein; that is, those spin configurations in which a single cell is occupied at each sequence number slice. The connectivity of the protein can also be assured by constraining length of bonds between neighboring sequence number slices. For such configurations, the lattice is merely an artifice. One can write the energy function in terms of the Cartesian coordinates of the α-carbons:

$$H = -\sum_{i \neq j} \sum_{m,n} \lambda_{m,n} h_{m,i} h_{n,\delta} h^\alpha_{m,i} h^\alpha_{n,j} \theta_{SW}(r_{ij} - r^\alpha_{ij}). \qquad (40)$$

In this equation, the h_m's are the charges of type m associated with the protein to be folded; the h_α, the charges in the proteins found in the data base. That is, θ_{SW} is a square well function of the distance between the sites, centered about zero in each of the proteins in the data base. Because this Hamiltonian is a function of only scalar distances, it is clearly invariant with respect to translation and rotation. Unlike the usual potentials inferred from physical arguments, the interactions in this Hamiltonian are very long range and mean field theories are likely to be very accurate for them. Thus the results which we discussed in the last section, based on the statistical Flory theory, can give us a good idea of the qualitative nature of the phase diagram. The interactions from the correct data base protein act like the ordering part of the Hamiltonian. The other memories give the random contributions. In many ways, the theory of associative memory polymer Hamiltonians parallels that of the Hopfield spin memory, with the exception that transitions behave more like those in the Potts glass because of the lack of Ising symmetries.

Structurally the associative memory Hamiltonians resemble the Mattis models of Garel and Orland, but the long-range nature of the interactions means the folding translation is like crystallization, *not* simple collapse. These Hamiltonians will use data base proteins that are likely to be correlated. Certainly one of the major issues is the appropriate weight to be assigned memories. In any event, the correlations is likely to affect favorably the ability of the associative memory model to recall (but not differentiate) structures, because the correlations can act to reinforce each other constructively rather than destructively. If the correlations are to be appropriate, it is best to choose charge densities that have good biological and biophysical movitation. In trying to assess the usefulness of these associative memory Hamiltonians, we must rely upon simulation techniques because of the complex and correlated nature of the memories in the data base.

To begin the investigation, Friedrichs and Wolynes chose to characterize the sequence by hydrophobic charges. As we remarked in the earlier sections, the large scale structure of globulin proteins is determined in part by the arrangement of the hydrophobic residues. Since the local secondary structures are periodic as to interior and exterior, α-helices and β-sheets generally have characteristic repeating patterns of alternating hydrophobicity.[49] Such patterns have been used by Eisenberg to discuss the topology of membrane proteins.[50]

In addition to using the hydrophobic charge based on Eisenberg's consensus scale, it was useful to include the interaction with a mass charge that simply measures whether a residue is present or not. This changes the Hamiltonian to

$$H = -\sum_{i \neq j} \sum_{\alpha} (h_i h_j h_i^\alpha h_j^\alpha + h_i h_i^\alpha + h_j h_j^\alpha + \gamma) \theta_{\mathrm{SW}}(r_{ij} r_{ij}^\alpha). \qquad (41)$$

Notice that this additional term is very much like the terms added to improve the memory of sparse correlated patterns introduced in the theory of neural nets.[48]

The first question to be asked about such an associative memory Hamiltonian is, what is its capacity? Technically, this occurs when the number of memories in the data base is so large that the target is no longer near the minimum of a Hamiltonian. This can be roughly investigated by finding the energy cost of introducing defects into the target structure. This can be done in two ways: (1) by the deterministic checking of each single site; and (2) by the use of simulated annealing, constraining Monte Carlo moves to small deviations from the target. Such a study has been carried out using the protein Rubredoxin (a small β-sheet protein) as the target and with the use of data sets of memory proteins, including Rubredoxin, but also including ones uncorrelated with it. For data bases of order 40 and 80, this test of capacity shows that there would be very few errors at low temperature in such models. Perhaps more important than capacity is the question of accessibility. Can one start from at least a partially disordered structure and find the appropriate minimum? Friedrichs and Wolynes have approached this question again for Rubredoxin using simulated annealing from a high temperature structure. Since the transition is in the best case certain to involve nucleation, they studied this recall by using initial structures which have some of the residues frozen into place and allowing only the other fraction of the residues to move. An example of the kind of structure which can be obtained is shown in Fig. 1. Here the exact Rubredoxin structure was obtained from an anneal from a 40 memory data base with the first 20 residues held in place. The resolution of this structure is 1.7. Such a resolution would be quite acceptable as the starting point for an energy minimization using a more complete description of the protein molecule.

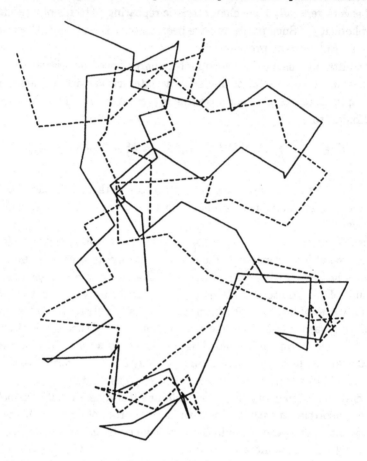

Fig. 1. Overlap of crystallographic "Clostridium Pasteurianium" rubredoxin (dotted line) with structure calculated with a 40 protein memory data base (solid line).

When larger data bases are used with this same nucleation size, lower resolution is obtained and this resolution is diminished when smaller nucleated sections are used. This is, of course, very much in keeping with the expectation of a model in which some glassy dynamics is occurring on the way to the quench. Considerably more work must be done to show that such associative memory Hamiltonian are predictive. The large capacity revealed by this study, although it is much less than the total number of proteins imaginable, is very encouraging because of the belief that the number of families of proteins is relatively small and of this order. Thus, an intelligently weighted sampling of the families of the data base may well lead to a useful way of classifying new sequences. Even if the associative memory models prove not to be powerful in predicting general protein structure, this capacity may well be sufficient for understanding the structures within a small group of proteins.

Perhaps the most important possibility here is the investigation of the variable regions of immunoglobulins which form a rather large class.[51] The issue of a finite capacity of such models is also interesting in that context because of the problems of the maximum size of the immune repertoire.

6. Questions Raised by the Spin Glass Picture in the Protein Folding Problem

The exploration of spin glass ideas in the biological problem of protein folding has only begun. Certainly the value of the spin glass approach is not in the problems which it has answered so much as the way in which it allows one to ask new questions. In this final section, I would like to raise, more explicitly than in the earlier text, some of these questions.

1. The first question, of course, is does it make any sense at all to use the ideas of statistical proteins to think about folding? Proteins are, after all, the result of many years of biological evolution. They are far from random.

   Only time will tell! However, the more precise way of asking this question is to wonder whether evolution has been primarily driven in designing sequences so that they will fold, or whether this is in some sense one of the least demanding constraints on biological evolution. In any event, the statistical protein and the associative memory models allow one to interpolate between a situation in which easy folding occurs to one in which it is extremely difficult.

2. Second, do spin glass phases occur in a natural protein? Can such phases be exploited in control mechanisms by organisms? In a sense, if a spin glass phase does occur, this would mean that misfolding is essential at equilibrium. It would then be necessary for the living system to provide some kinetic means of avoiding this misfolding. Biological mechanisms may exploit this opportunity. A relevant example occurs, not in protein folding, but in nucleic acid folding. One mechanism of controlling RNA

synthesis relies on the production of metastable structures.[52] If RNA is allowed to equilibrate it leads to misfolded structures. The equilibration process is tied to the presence of needed amino acids and, hence, this acts as a control function. Clearly this can happen more easily with nucleic acids because of their less compact structure and larger energies of association through base pairing. There is evidence that this may also occur in some situations with proteins as well, but considerable more work must be done on this.[2] It could well play a role in the biology of aging and degradation of proteins.

3. Can associative memory models be used to predict protein structures? The issues here will certainly involve the question of finding the appropriate invariances of associative memory Hamiltonians. These would include invariance to mutation of sites, as well as insertions and deletions. A great deal of work has been done in the theory of protein sequences that may give ideas about how to proceed.[63]

It will clearly also be necessary to understand how the capacity of associative memory Hamiltonians can be increased. Here many of the ideas from spin associative memories may prove useful. It is known there, for example, that many spin interactions can be helpful in increasing that capacity.[54] Many body interactions may play this same role here.

A further set of questions about the associative memory Hamiltonians arises from imagining how the accessibility of the correct final state can be enhanced. The theory of the dynamics of this models suggests this can be done by varying the Hamiltonians so that the ratio of the glass temperature to the folding temperature is small. Here the analytical theories of the glass transition in statistical proteins may be very helpful.

4. Are there better optimizing algorithms? Associative memory Hamiltonian and other statistical Hamiltonians may provide a good laboratory for investigating new optimization techniques very different from simulated annealing. For example, the genetic algorithms will be interesting in the context of those models for which the difficulty of simulated annealing can be explicitly controlled.[55,56] The associative memory Hamiltonian can be changed, rather than the temperature, as the optimization goes along. (Furthermore, since the associative memory Hamiltonian is, in some sense, a device, one may imagine using the theory of the spin glass temperatures to design optimal annealing schedules.)

5. Does the natural organization of spin ground states into families have any connection with the evolutionary process? Does it place constraints on the mechanism of molecular recognition, i.e., "lock and key"[57] or "induced fit" models[58] for catalysis, or the nature of allosteric transitions?

6. Does the glassy dynamics of folded proteins have any connection with folding intermediates? Frauenfelder has recently argued that the highest tiers of organization of conformational substates of myoglobin may be partially misfolded structures.[59]

In addition to raising these biophysical questions, the connection of the spin glass problem with the protein folding question will act as a major incentive to investigate further the properties of more complex spin glass models. The role of dilution and the use of short-range interactions are problems that immediately occur in thinking about real proteins, and the study of these in the context of the folding problem may help in generally fleshing out those areas. Better knowledge of these areas may well pay off for the other problems that are discussed in this volume.

Acknowledgments

I have enjoyed and benefitted from my interactions with many people in connected with the folding problem. I thank R. Baldwin, J. D. Bryngelson, H. Drickamer, H. Frauenfelder, M. Friedrichs, P. Goldbart, S. Kaufman, T. Kirkpatrick, J. McCammon, E. Shakhnovich, L. Smarr, D. Stein, K. Schulten, Z. Schulten, G. Weber, J. Widom, and R. Zwanzig. My work on folding has been supported by grants NSF CHE84-18619 and NSF DMR 86-12860. I also thank Betty Brillhart for cheerfully typing the manuscript, under much pressure.

References

1. C. Anfinsen, *Science* **181**, 223 (1973).
2. G. C. Flyn, T. G. Chappell, and J. E. Rothman, *Science* **245**, 385 (1989); G. Fisher *et al.*, *Biomed. Biochim. Meth.* **43**, 1101 (1984); A. Townsend *et al.*, *Nature* **340**, 443 (1989).
3. a) J. Hopfield, *Proc. Natl. Acad. Sci.* **79**, 2554 (1984).
   b) J. Hopfield and D. W. Tank, *Science* **233**, 625 (1985).
4. S. A. Kauffman, E. Weinberger, and A. S. Perelson, in *Theoretical Immunology*, Part One, SFI Studies in the Science of Complexity, ed. A. S. Perelson (Addison-Wesley, 1988), pp. 349–382.
5. P. W. Anderson, *Proc. Natl. Acad. Sci.* **80**, 3386 (1983).
6. One excellent book on protein structure is G. E. Schulz and R. H. Schimer,

Principles of Protein Structure (Springer, 1979). A book on the specific protein hemoglobin is also a wonderful place to start learning about protein structure: R. E. Dickerson and I. Geis, *Hemoglobin: Structure, Function, Evolution and Pathology* (Benjamin, 1983).

7. F. C. Chothia, M. Levitt, and D. Richardson, *Proc. Natl. Acad. Sci.* **73**, 3793 (1977).

8. F. Crick, *Acta, Cryst.* **6**, 689 (1953).

9. J. Richardson, *Adv. Protein Chem.* **34**, 167 (1981).

10. a) Yu. N. Chirgadze, *Acta Cryst.* **A43**, 405 (1987).
    b) A. G. Murzin and A. V. Finkelstein, *J. Mol. Biol.* **204**, 749 (1988).

11. D. E. Wetlaufer, *Proc. Natl. Acad. Sci.* **70**, 697 (1973).

12. A comprehensive review of the experimental literature until 1980 is C. Ghelis and J. Yon, *Protein Folding* (Academic, 1982). Another useful review is P. S. Lim and R. L. Baldwin, *Ann. Rev. Biochem.* **51** 459 (1982).

13. P. L. Privalov, *Ann. Rev. Biophys. Biophys. Chem.* **18**, 47 (1989).

14. D. A. Dolgikh *et al.*, *FEBS Lett.* **165**, 88 (1984); G. Damuschin *et al., Int. J. Biol. Macromol.* **8** 226 (1986).

15. T. E. Creighton, *J. Mol. Biol.* **87**, 563, 579, 603 (1974).

16. See the discussion in Ghelis and Yon on elastase. Similarly for Phosphofructokinase, see G. R. Parr and G. Hammes, *Biochem.* **14**, 1600 (1975).

17. J. D. Bryngelson and P. G. Wolynes, *Proc. Natl. Acad. Sci.* **84**, 7524 (1987).

18. P. J. Flory, *Principles of Polymer Chemistry* (Cornell, 1953).

19. a) J. Deutch and H. Hentschel, *J. Chem. Phys.* **85**, 527 (1986).
    b) I. Lifshitz, A. Grosberg and A. Khokhlov, *Rev. Mod. Phys.* **50**, 573 (1978).

20. D. Poland and H. Scherage, *Theory of Helix Coil Transitions in Biopolymers* (Academic, 1970).

21. The formation of β sheets does in fact correspond with a sharp phase transition. It is described by model which can be solved exactly: R. Zwanzig and J. I. Lauritsen, *J. Chem. Phys.* **48**, 3351 (1968).

22. J. D. Bryngelson and P. G. Wolynes, *Biopolymers* **30**, 177 (1990).

23. B. Derrida, *Phys. Rep.* **67**, 29 (1980); *Phys. Rev. Lett.* **45**, 79 (1980); D. Gross and M. Mezard, *Nucl. Phys.* **B240**, 431 (1984).

24. The REA has been used in describing glass transitions in folded polymers in a nice paper by D. L. Stein, *Proc. Natl. Acad. Sci.* **82**, 3670 (1985).

25. A. Ansaris *et al.*, *Proc. Natl. Acad. Sci.* **82**, 5000 (1985).

26. A. Ansari *et al.*, *Biophys. Chem.* **26**, 337 (1987); I. E. T. Iben *et al., Phys. Rev. Lett.* **62**, 1916 (1989).

27. J. D. Bryngelson and P. G. Wolynes, *J. Phys. Chem.* **93**, 6902 (1989).

28. A. Baumgärtner, *Ann. Rev. Phys. Chem.* **35**, 419 (1984).

29. In a different context this result was obtained by S. Kauffman and S. Levin, *J. Theor. Biol.* **128**, 11 (1987).

30. One-dimensional random potentials give similar results. See H. Bässler, *Phys. Rev. Lett.* **58**, 767 (1987) and R. Zwanzig, *Proc. Natl. Acad. Sci.* **85**, 2029 (1988).

31. M. Shlesinger, *Ann. Rev. Phys. Chem.* **39**, 269 (1988).
32. A. Szabo, Z. Schulten, and K. Schulten, *J. Chem. Phys.* **72**, 4350 (1980).
33. J. M. Deutch, *J. Chem. Phys.* **73**, 4700 (1980).
34. C. Levinthal, *J. Chem. Phys.* **65**, 44 (1968).
35. This is very much in keeping with the arguments of S. Harrison and R. Durbin, *Proc. Natl. Acad. Sci.* **82**, 4028 (1985).
36. T. Garel and H. Orland, *Europhys. Lett.* **6**, 307 (1988).
37. T. Garel and H. Orland, *Europhys. Lett.* **6**, 597 (1988).
38. E. I. Shakhnovich and A. M. Gutin, *Europhys. Lett.* **8**, 327 (1989).
39. E. I. Shakhnovich and A. M. Gutin, *J. Phys.* **A22**, 1647 (1989).
40. E. I. Shakhnovich and A. M. Gutin, *Biophys. Chem.*, to appear.
41. T. R. Kirkpatrick and P. G. Wolynes, *Phys. Rev.* **B36**, 8552 (1987).
42. H. Frauenfelder, F. Parak, and R. D. Young, *Ann. Rev. Biophys. Biophys. Chem.* **17**, 451 (1988).
43. P. de Gennes, *Scaling Concepts in Polymer Physics* (Cornell, 1979).
44. P. Golbart and N. Goldenfeld, *Phys. Rev. Lett.* **58**, 2676 (1987); *Phys. Rev.* **A39**, 1402 (1989); *ibid.* **39**, 1412 (1989); *Macromolecules* **22**, 948 (1989).
45. Paul Goldbart, private communication.
46. W. Braun, *Quart. Rev. Biophys.* **19**, 115 (1987).
47. M. Friedrichs and P. C. Wolynes, *Science* **246**, 371 (1989).
48. L. Personnaz, I. Guyon and G. Dreyfus, *J. Phys. Lett.* **46**, L359 (1985); J. P. Nadal *et al.*, *Europhys. Lett.* **1**, 535 (1986); I. Kanter and H. Sompolinsky, *Phys. Rev.* **A35**, 380 (1987).
49. M. Shiffer and A. B. Edmundson, *Biophys. J.* **7**, 121 (1967); J. Palau and P. Puigdomenech, *J. Mol. Biol.* **88**, 457 (1974); V. I. Lin, *J. Mol. Biol.* **88**, 857 (1974); V. I. Lim, *J. Mol. Biol.* **88**, 857 (1974).
50. D. Eisenberg *et al.*, *Faraday Symp. Chem. Soc.* **17**, 109 (1982); D. Eisenberg *et al.*, *Proc. Natl. Acad. Sci.* **81**, 140 (1984).
51. R. M. Alzari *et al.*, *Ann. Rev. Immunol.* **6**, 555 (1980); D. R. Davies *et al.*, *J. Biol. Chem.* **263**, 10541 (1988).
52. A. Das, I. P. Crawford and C. Yanofsky, *J. Biol. Chem.* **257**, 8795 (1982).
53. G. V. Heijne, *Sequence Analysis in Molecular Biology* (Academic, 1987).
54. L. F. Abbott and Y. Arian, *Phys. Rev.* **A36**, 5091 (1987).
55. D. E. Goldberg, *Genetic Algorithms in Search, Optimization and Machine Learning* (Addison-Wesley, 1989).
56. M. Friedrichs and P. G. Wolynes, preprint.
57. E. Fisher, *Ber. Chem. Ges.* **27**, 2985 (1984).
58. D. E. Koshland, *Proc. Natl. Acad. Sci.* **44**, 98 (1958).
59. H. Frauenfelder *et al.*, contribution to H. G. Drickamer Festschrift.